Lecture Notes in Social Networks

More information about this series at http://www.springer.com/series/8768

Mehmet Kaya • Şuayip Birinci • Jalal Kawash •
Reda Alhajj
Editors

Putting Social Media and Networking Data in Practice for Education, Planning, Prediction and Recommendation

Editors
Mehmet Kaya
Computer Engineering
Fırat University
Elazığ, Turkey

Şuayip Birinci
Ministry of Health
Ankara, Turkey

Jalal Kawash
Department of Computer Science
University of Calgary
Calgary, AB, Canada

Reda Alhajj
Department of Computer Science
University of Calgary
Calgary, AB, Canada

ISSN 2190-5428 ISSN 2190-5436 (electronic)
Lecture Notes in Social Networks
ISBN 978-3-030-33700-1 ISBN 978-3-030-33698-1 (eBook)
https://doi.org/10.1007/978-3-030-33698-1

This Springer imprint is published by the registered company Springer Nature Switzerland AG.
The registered company address is: Gewerbestrasse 11, 6330 Cham, Switzerland

Preface

Social networks (SN) have brought an unprecedented revolution affecting how people interact and socialize. SN have invaded almost all domains: lifestyle, medicine, business, education, politics, activism, and more. The result is billions of SN users. For example, Twitter claimed to have 321 million monthly active users in 2018, and Facebook had 2.41 billion monthly active users as of June 2019. Online social media (OSM), media produced by SN users, have offered a real and viable alternative to conventional mainstream media. OSM are likely to provide "raw," unedited information and the details can be overwhelming with the potential of misinformation and disinformation. Despite these dangers, OSM are leading to the democratization of knowledge and information. OSM are allowing almost any citizen to become a journalist reporting on specific events of interest. This is resulting in unimaginable amounts of information being shared among huge numbers of OSM participants. For example, Facebook users are generating several billion "likes" and more than 100 million posted pictures in a single day. Twitter users are producing more than 6000 tweets per second. The size of the data generated presents increasing challenges to mine, analyze, utilize, and exploit such content. For example, analyzing the prohibitively long clickstreams in e-commerce applications and sifting through an overwhelming number of research articles become very challenging. At the same time, this explosion is opening doors to new exploitation and application of the generated data. For instance, various recommendation systems can make use of these data mines to generate more accurate and relevant recommendations.

This book includes twelve contributions that examine several topics related to the utilization and exploitation of SN and OSM. The topics are emergency evacuation scenarios, creating research groups, recommendation for food venues, clickstream analysis for e-commerce, event detection, fraud processes, scientific article recommendation, popular vs. unpopular activities, alternative navigation route recommendation, sentiment analysis, clickbait analysis, and the detection of anomalies.

Modeling crowd behavior in emergency evacuation scenarios is the subject of a chapter by Sahin and Alhajj. Using a belief-desire-intention model, they propose a multiagent system, which works with partially observable environments by the agents. Individuals perceive partial information about the environment and other evacuees continuously, and a belief set is created accordingly. A preconstructed set of plans is filtered using the current beliefs to find subtasks. Test scenarios and reported results are encouraging.

Green et al. report on partnering with students to perform a research project that investigated supporting students transitioning from first year to a Bachelor of Primary Education program. Students and their academic mentor were connected through a Facebook group, whose aim is to provide support to the group of transitioning students. Utilizing social media, students could access support from their peers and university staff and generate a community of learners.

Recommending feed venues using multilingual social media content is the subject of a chapter by Siriaraya et al. The system makes use of region-based popularity of venues, user rating, and tweet sentiment analysis in order to generate the recommendations. The system is experimentally validated using 26 million tweets from different European countries for four different recommendation approaches.

Xylogiannopoulos et al. embark on simplifying e-commerce analytics by discovering hidden knowledge in big data clickstreams. The massive number of the combination of online users, retailers, and products makes clickstream analysis very challenging. The authors address this issue by significantly modifying and upgrading a sequential frequent itemsets detection methodology. The result is a method that can do the analysis using very limited computational power, such as a desktop computer. They demonstrate the efficiency of their methods using 10 billion records related to top US online retailers.

Event detection in communities extracted from communication networks is the subject of a chapter by Aktunc et al. They focus on social interactions in communities to detect events; specifically they focus on tracking the change in community structure within temporal communication networks. Many versions of the community change detection methods are developed using different models. Empirical analysis shows that community change can be used as an indicator of events, and the ensemble model further improves the event detection performance.

Thovex looks at chasing undetected fraud processes. Employing deep probabilistic networks, hidden fraud activities can be detected. Inspired by waves temporal reversal in finite space, offline analysis and mining of big hidden data are tackled. Experiments are encouraging and the proposed model may introduce an alternative in artificial intelligence for a new generation of applications.

Betül Bulut et al. tackle the problem of recommending relevant scientific articles to researchers. Finding good matches is becoming more challenging with the increasing number of publications. An article recommendation system is presented in this chapter. This system takes into consideration the researcher's previous downloaded articles and the field of study. As a result of the study, the proposed method has achieved more successful results compared to other existing methods.

Patterns of behavior of individuals and groups in community-driven discussion platforms, such as Reddit, is the subject of a chapter by Thukral et al. Using statistical analysis, they provide interesting insights about a large number of posts that failed to attract other users' attention despite their author's behavior, they provide more insights about short-lived but highly active posts, and they analyze controversial posts. Conclusions include understanding how social media evolves with time.

Using Google maps history logs to mine habitual user choices is the subject of a chapter by Varlamis et al. GPS logs are analyzed for durations of stay in certain locations and most frequent routes taken by the user. Using a trajectory partitioning method, they identify the most frequent subtrajectories followed by the user. This in turn allows recommendation for alternative travel routes, improving the user experience.

The subject of sentiment analysis is covered in a chapter by Martin-Gutierrez et al. Due to the limits of standard sentiment analysis (such as dependence on the size and quality of the training set and prelabeling), the chapter proposes a new methodology to retrieve text samples from Twitter and automatically label them. Experimental analysis applied to various Twitter conversations yields promising results.

Geçkil et al. study "clickbaits," a mechanism to manipulate a reader to click on links that lead to unwanted or irrelevant content. They look at how such baits can be detected on news sites. In this study, the data was gathered from news websites of media organizations and Twitter to detect clickbaits. The method employed yields a high accuracy rate of detection.

Bouguessa takes on the problem of identifying and automatically detecting anomalous nodes in networks. Nodes are represented with feature vectors, representing neighborhood connectivity. A probabilistic framework that uses the Dirichlet distribution is employed to anomalies. The result is an automatic discrimination between normal and anomalous nodes, rather than providing a ranked list of nodes and delegating the detection to the user. Experiments on both synthesized and real network data illustrate the suitability of the proposed method.

To conclude this preface, we would like to thank the authors who submitted papers and the reviewers who provided detailed constructive reports which improved the quality of the papers. Various people from Springer deserve large credit for their help and support in all the issues related to publishing this book.

Elazığ, Turkey — Mehmet Kaya
Ankara, Turkey — Şuayip Birinci
Calgary, AB, Canada — Jalal Kawash
Calgary, AB, Canada — Reda Alhajj
September 2019

Contents

Contributors

Aman Agarwal IIIT Delhi, New Delhi, India

Toyokazu Akiyama Kyoto Sangyo Daigaku, Kyoto, Japan

Riza Aktunc METU, Computer Engineering Department, Ankara, Turkey

Reda Alhajj Department of Computer Science, University of Calgary, Calgary, AB, Canada

Department of Computer Engineering, Istanbul Medipol University, Istanbul, Turkey

R. M. Benito Grupo de Sistemas Complejos, Escuela Técnica Superior de Ingeniería Agronómica, Alimentaria y de Biosistemas, Universidad Politécnica de Madrid, Madrid, Spain

Mohamed Bouguessa Department of Computer Science, University of Quebec at Montreal, Montreal, QC, Canada

Grigoris Bouras Department of Informatics and Telematics, Harokopio University of Athens, Athens, Greece

Betül Bulut Department of Computer Engineering, Fırat University, Elazığ, Turkey

Arnab Chatterjee TCS Research, New Delhi, India

Lipika Dey TCS Research, New Delhi, India

Michelle J. Eady University of Wollongong, Wollongong, NSW, Australia

Ayşe Geçkil Department of Computer Engineering, Fırat University, Elazığ, Turkey

Corinne A. Green University of Wollongong, Wollongong, NSW, Australia

Esra Gündoğan Department of Computer Engineering, Fırat University, Elazığ, Turkey

Pinar Karagoz METU, Computer Engineering Department, Ankara, Turkey

Panagiotis Karampelas Department of Informatics and Computers, Hellenic Air Force Academy, Dekelia Air Base, Acharnes, Greece

Tushar Kataria IIIT Delhi, New Delhi, India

Yukiko Kawai Kyoto Sangyo Daigaku, Kyoto, Japan

Buket Kaya Department of Electronics and Automation, Fırat University, Elazığ, Turkey

Mehmet Kaya Department of Computer Engineering, Fırat University, Elazığ, Turkey

J. C. Losada Grupo de Sistemas Complejos, Escuela Técnica Superior de Ingeniería Agronómica, Alimentaria y de Biosistemas, Universidad Politécnica de Madrid, Madrid, Spain

S. Martin-Gutierrez Grupo de Sistemas Complejos, Escuela Técnica Superior de Ingeniería Agronómica, Alimentaria y de Biosistemas, Universidad Politécnica de Madrid, Madrid, Spain

Emily McMillan University of Wollongong, Wollongong, NSW, Australia

Hardik Meisheri TCS Research, New Delhi, India

Ahmet Anıl Müngen Department of Computer Engineering, Fırat University, Elazığ, Turkey

Lachlan Munn University of Wollongong, Wollongong, NSW, Australia

Yusuke Nakaoka Kyoto Sangyo Daigaku, Kyoto, Japan

Coşkun Şahin Department of Computer Science, University of Calgary, Calgary, AB, Canada

Christos Sardianos Department of Informatics and Telematics, Harokopio University of Athens, Athens, Greece

Panote Siriaraya Kyoto Institute of Technology, Kyoto, Japan

Caitlin Sole University of Wollongong, Wollongong, NSW, Australia

Christophe Thovex DATA2B, Cesson Sevigné, France

Laboratoire Franco-Mexicain d'Informatique et d'Automatique, UMI CNRS 3175 – CINVESTAV, Zacatenco, Mexico D.F., Mexico

Sachin Thukral TCS Research, New Delhi, India

Ismail Hakki Toroslu METU, Computer Engineering Department, Ankara, Turkey

Iraklis Varlamis Department of Informatics and Telematics, Harokopio University of Athens, Athens, Greece

Ishan Verma TCS Research, New Delhi, India

Yuanyuan Wang Yamaguchi Daigaku, Ube, Japan

Konstantinos F. Xylogiannopoulos Department of Computer Science, University of Calgary, Calgary, AB, Canada

Crowd Behavior Modeling in Emergency Evacuation Scenarios Using Belief-Desire-Intention Model

Coşkun Şahin and Reda Alhajj

1 Introduction

Realistic crowd behavior modeling has received considerable attention from the research community over the last two decades. Researchers applied different methods to imitate human actions for specific cases. Some of these works create cognitive models which focus on psychological aspects, while others consider physical forces as the main component of the whole process. There are works which considered the other evacuees as the major factor during the decision making process. Thus, they built the system based on the interactions among individuals and by considering group behavior models. These interactions are affected by various factors, including ethnicity, personal risk, credibility, social status, emotional status, personality, etc. [1, 2].

As it is hard to model crowd behavior by using a single source of elements, hybrid solutions with multiple components are becoming more popular and attractive to reflect more realistic scenarios. However, the set of factors affecting the decisions of an individual is huge. Especially during emergency situations, there are more aspects to take into consideration. For instance, panic, fear and chaos may prevent evacuees from acting reasonably and may lead to unforeseen scenarios. Thus, realistic behavior modeling in these cases is mostly different from strategic planning.

C. Şahin (✉)
Department of Computer Science, University of Calgary, Calgary, AB, Canada
e-mail: coskun.sahin1@ucalgary.ca

R. Alhajj
Department of Computer Science, University of Calgary, Calgary, AB, Canada

Department of Computer Engineering, Istanbul Medipol University, Istanbul, Turkey
e-mail: alhajj@ucalgary.ca

M. Kaya et al. (eds.), *Putting Social Media and Networking Data in Practice for Education, Planning, Prediction and Recommendation*, Lecture Notes in Social Networks, https://doi.org/10.1007/978-3-030-33698-1_1

Another important aspect in realistic crowd behavior modeling is incorporating fuzziness in the framework. In real-time situations, taking an action may not lead to the intended results. Also, there may be some accidents as a result of being careless or due to false assumptions. For instance, it is possible to see people colliding at corners or falling from the stairs after missing a step. However, an extremely detailed cognitive and physical model would fail in producing real-time output. Because considering every single detail for every individual in a continuous decision making process cannot be handled in a restricted amount of time with limited resources. As the number of factors affecting an individual is enormously big, some behavior without obvious reasons can be modeled as a part of randomness. For example, seeing an evacuee running in the opposite direction may make the others scared so that they unintentionally start to do the same thing. They may follow him/her even if they do not know the actual reason for running in the opposite direction. Similarly, misinterpretation of observations could result in taking faulty actions.

Partial observability is another crucial part of a realistic simulation model because, in real world scenarios, perceptions of the evacuees may not be clear or easy to comprehend. Every type of input, including smell, visual data, sound and physical contact is received independently. It is the job of the individual to evaluate and relate them with each other. By combining different inputs, a human-specific set of beliefs is created. There could be many irrelevant data, such as noise from other humans, which should be ignored. Moreover, it is possible to retrieve misleading data. For instance, hearing something incorrectly from another person or any type of misunderstanding is not an unusual event, especially in a crowded environment.

Similar to the cognitive process, perceptions coming to a simulation agent should be treated as independent data and the agent should be responsible for gathering them together to form something meaningful. Restricting perception range for vision/sound and making the agent find its own way using the data received during exploration could lead to a better crowd behavior model. However, it is important to consider the idea that each individual has limited resources and every single data cannot be processed in a detailed way. This phenomenon is known as bounded rationality [3].

As a result of the points explained above, creating a hybrid model containing the crucial aspects of real world planning is the key for realistic agent modeling. The model should represent processes such as perception, cognition, goal seeking, task scheduling, collaboration and communication [4].

The aim of the work described in this chapter is to create a partially observable environment and agent model as part of our existing multi-agent emergency evacuation simulation system. Among different approaches, we selected the Belief-Desire-Intention (BDI) framework which provides a high-level planning mechanism. Thus, it allows for the abstraction of low-level individual tasks, such as going from one location to another. The following section discusses the reasons of selecting BDI instead of some other approaches. Actually, the results reported in this chapter demonstrate the effectiveness and applicability of this approach for real-world scenarios.

The rest of this chapter is organized as follows. Section 2 gives background information and some example systems with their adapted approaches. Section 3 covers the developed emergency detection system and integration using the evacuation simulation system. In Sect. 4, we discuss the proposed BDI model and we describe the implementation details. Example simulations and result analysis of the system are provided in Sect. 5. Finally, Sect. 6 is conclusions and future work.

2 Background and Related Work

Agent-based architectures are commonly used in crowd behavior modeling. The main reason is the flexibility to represent each agent by its own attribute set with different values. This leads researchers to create heterogeneous environments where agents perform actions independently. Moreover, the architecture gives designers the flexibility to model the interaction among individuals and to simulate the group behavior. Thus, it is possible to use both microscopic and macroscopic approaches.

Dawson et al. [5] used an agent-based simulation approach to analyze the effects of flooding on people and provided a risk-based flood incident management (FIM) model. The responses of an agent are determined based on the hydrodynamic and the other environmental changes during a flooding incident. They prefer a probabilistic finite state machine to model states, actions and transitions. Some other works also used a multi-agent model with a feature set, e.g., [6–9].

There are various techniques adopted for modeling the decision making process of an individual. Partially-observable Markov decision processes (POMDP), the Belief-Desire-Intention (BDI) [10] framework, and cognitive architectures have been widely used for different purposes in the past. The BDI framework basically depends on the philosophical explanation of how humans make decisions. The details of performing actions are omitted and the main focus is on the process itself. Cognitive architectures aim to model the way human brain works as precisely as possible. Thus, the process depends on the results of neurological studies conducted on cognitive processes. Another difference between these two approaches is their abstraction level. Cognitive methods are low-level formulations of problems while the BDI model consists of mental steps of taking the action; it ignores the other parts. As the technical details are omitted, it is easier for non-computer science researchers to understand the steps of the BDI model.

POMDPs are used to model a learning problem in a structured way. Differently from a classical MDP structure, an agent in a POMDP environment cannot directly observe the current state. Thus, it uses a probabilistic approach on its observations. Various methods which have been used for solving MDPs, such as policy iterations and value iteration algorithms, have also been adapted for solving POMDPs, as well. The main objective in these approaches is to find the optimal solutions which lead to the maximum reward in the environment.

The POMDP framework is an effective way of representing some learning problems. The reward mechanism provides feedback to the agent during the process.

However, finding optimal POMDP policies is usually intractable. In addition, it is not easy to define multiple major tasks or some optional sub-tasks. On the other hand, the BDI model does not have a rewarding mechanism, but it is practical to introduce multiple tasks. Its plan-set is not adaptive, and it is not guaranteed to provide optimal solutions within a changing environment. Thus, both approaches have advantages and drawbacks. Accordingly, hybrid models may be more desirable for specific cases [11–13].

In addition to emergency evacuation, adaptive intelligent agents are needed in various types of problems, especially for crowd modeling. Video games, business process simulations and animal colony simulations are some of the applications of adaptive agents in dynamic environments. An agent-based BDI model is one of the most common choices for this purpose. In addition to early popular applications, such as Procedural Reasoning System (PRS) [14] and dMARS [15], the BDI approach has been adopted in some other works. It is not uncommon to see some hybrid solutions where the BDI framework cooperates or is combined with other techniques.

Zhao et al. [16] utilized the agent based BDI model for an automated manufacturing system. Their agent model is in the role of an operator who is responsible for detecting errors in an automated shop floor control system. The traditional BDI framework has been extended to include a deliberator, a planner and a decision executor. They also introduced the confidence index of a human which indicates how successful the agent has been with its latest actions. In case of poor performance, it reconsiders every intention before applying actions. The proposed model is adaptive and dynamically changes its behavior with the changing environment.

Lee et al. [17] used a similar modified BDI framework for mimicking realistic human behavior. In addition to BDI, they utilized the Bayesian Belief Network (BBN), the Decision Field Theory (DFT) and Probabilistic Depth First Search (PDFS) for planning and performing actions. The BBN model is employed to capture probabilistic relationships and historical data about the environment. The DFT model provides mathematical representation of the preferences of agents during the decision making process. It is used to calculate the preference values of each option. PDFS is the plan selection mechanism based on the preference values. It generates every option for the current situation and searches for a suitable one in a depth first search manner. The authors applied the proposed model for an evacuation scenario of a terrorist bomb attack.

Okaya et al. [2] also integrated the BDI model with a different method. They combined Helbing's social force model [18] with an agent-based BDI structure to simulate physical forces, psychological states, environment knowledge of agents and inter-agent communication at the same time. The agents are modeled to receive visual and auditory input from the environment. Perceptions are used to generate human relation factors created as a union of personal risks, family context and adaptive plans. Civilians are created as part of one of three groups, namely an adult, a parent or a child. Depending on their types, their attitude towards the others and their collaborative abilities are formed. Social forces on an agent are calculated using Helbing's model and they are modified by the intentions of an agent and its attribute values.

Liu et al. [19] used navigational agents to group individuals in an evacuation area. Their model contains a knowledge-base with different sets of information. They created a two-layer control mechanism with a social force model and a cultural algorithm (CA) [20], which has been designed for modeling the cultural evolution process. The approach is effective in terms of dividing the crowd behavior simulation problem into sub-groups each of which is guided by a leader.

Adam et al. [21] studied Melbourne bush fire in 2009 by building two models based on a Finite State Machine (FSM) and the BDI framework. By addressing the same problem with different architectures, they compared the models in terms of their effectiveness. They argued the following, while FSM requires generating every possible state and the transitions between each other, the BDI model is more convenient for designing reactive agents. However, the BDI model is a more complex structure and may require greater computational power. Their comparison is based on multiple factors, including the statistics of the real incident and the simulated ones.

For some public events such as concerts, sports competitions and public places such as airports, majority of the individuals may be unfamiliar with the building. However, in an office where people see each other daily, relations among people and the fact that people know the environment cannot be ignored during egress. Many of the recent works rely on this fact, including the work of Valette et al. [22]. They created a heterogeneous environment of agents with social skills and emotions. Decision making is performed by the BDI model which also considers the attributes of an agent during the process. The proposed model has been applied to multiple real-world scenarios in order to evaluate its reliability.

3 Multi-Agent Emergency Evacuation Simulation Model

Our emergency evacuation simulation system is an agent-based model which is intended to be used for running egress simulations. The main objective is creating evacuation test case scenarios which are dangerous or impractical to build in real life. It is crucial to analyze every aspect of a building for emergency management purposes, even before starting the construction. In addition, this may be helpful for seating arrangements and exploring capacity limitations of buildings in case of emergencies.

The system uses a set of physical, mental and social attributes to model different types of humans in an evacuation process. Our primary concern in this part of the evacuation planning project is to model individuals and groups as realistic as possible. This way, different scenarios can be run and analyzed correctly. We mainly focus on crowd modeling and human decision making process. We leave the low-level aspects of physical interactions to the Unity game engine.

We use position, health, mobility, gender, age, stress level, panic level and social role attributes for each agent. Social role determines if the agent is a leader for the group or a follower. Depending on the physical abilities, an agent can choose resting, moving to a specific position or following a group/leader. Speeds of agents are calculated by a fuzzy logic engine which uses physical and mental features in fuzzy inference rules.

The model validation is done by conducting some experiments and observing particular individual and group behavior, such as selfish and collaborative attitudes. In addition, common phenomena like herding, arching and clogging are also detected. The model has been tested on a building floor with different numbers of evacuees while varying exit gate widths.

This architecture lacks of partial observability and agent-to-agent communication capabilities. In real world, especially during panic situations, it is not easy to interpret every input data correctly and in a fast way. Moreover, it is not a good idea for each agent to assume full knowledge about the building plan. Thus, in this work, our aim is to introduce a multi-agent BDI framework, which models individual planning and agent interactions in a more realistic way.

4 BDI Framework in Emergency Evacuation

4.1 BDI Framework

BDI framework is a common approach for building multi-agent systems where a human behavior model is necessary. It adopts the human decision making psychology by using beliefs, desires and intentions. Beliefs are the combination observational data with the information retrieved from other agents. Desires are the main goals of an agent. They tend to stay the same during the process, unless something unexpected occurs. Extraordinary cases could result in changes int the objective of an agent. In an emergency evacuation, for instance, the desire is leaving the building. However, a mother would mostly prefer saving her children, an action overrides the current desire. Similarly, for an agent with partial knowledge, desire may be expressed as reaching to a particular exit. However, if the pathway is blocked by obstacles or other evacuees, it may need to define a new desire. Intentions are simple actions or small checkpoints to achieve the final goals. For example, in order to leave a building, the first step for an agent could be exiting from the current room. Thus, it needs a partial plan to reach to the room door while staying away from obstacles and other evacuees. Typically, a BDI framework contains this type of predefined sub-plans which can be adopted by agents.

The need for a BDI framework arises because of the issues related to simple environment models, ignoring some crucial factors which exist in real world. First of all, in real world cases, the environment is highly dynamic. Possible presence of fire, smoke or flood and the other individuals performing actions should be considered continuously. A plan which is built based on the current snapshot of the environment could be obsolete while performing related actions. For example, in case of clogging in one of the doors, agent will not be able to continue. Thus, it should adapt its behavior according to the changes. Secondly, the amount of resources an individual can use is limited. Therefore, it may need to ignore some past information or irrelevant up-to-date data while deciding on next action. In addition, some perceptions could be noisy. This is the reason for calling any input as *beliefs* instead of *facts*. Thus, relying some observations for a long time could be harmful in case they are somehow misinterpreted.

4.2 Proposed Model

This part of the project aims to integrate a custom BDI framework to our previous work described in Sect. 3. The main reason is building a partially observable multi-agent model capable of adapting agents' behavior according to their perception. This type of approach is effective mainly when the state of the environment constantly changes. In addition, it will allow us to imitate psychological steps of processing input, using it for decision making and performing actions. The BDI architecture works with the existing modules in every part of the process. The proposed model is depicted in Fig. 1.

The architecture consists of a set of components which process, filter and prepare data for the other components. The main source of data is the environment itself. Each agent constantly collects information as *perception*. The major components of the system are:

(1) Input Manager Each agent continuously collects data from the environment via its sensors. Even though it is possible to process any kind of data in the following steps, the current setup only considers visual data. Ideally, every data type should include its propagation dynamics. For instance, in order to simulate smelling smoke, feeling fire, hearing the sound of other agents and the motion of flooding, additional dynamics should be defined. For visual data, the technique we use is sending multiple rays to different directions in the vision range of the agent and detecting the first object that the corresponding ray intersects with. This way, agents detect every obstacle, evacuee and gate seen in its specific vision range. Figure 2 shows two agents collecting visual data in a test environment. These objects are stored in the *input queue*. In addition to direct object detection, agents are capable of detecting components of the building and their connections. This provides locational awareness to agents. For instance, the bottom agent in Fig. 2 can detect that it is in *Room 2* and there are two close-by doors, namely *Door 1* and *Door2*. Initially, it does not know anything about the presence of neither the other agent nor the other building components. Thus input manager provides the information of what is around and where the agent in terms of belief. The filtered perceptions are passed to the *beliefs* list.

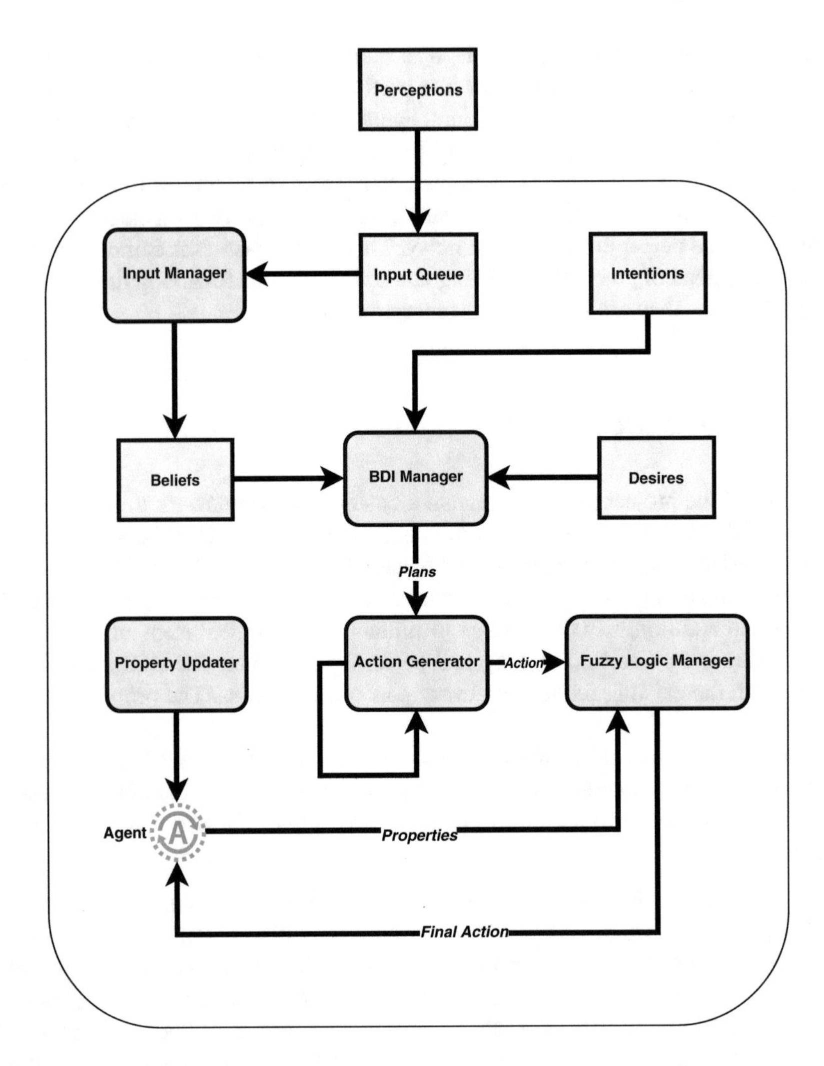

Fig. 1 Integrated BDI architecture

(2) BDI Manager BDI Manager uses the belief, desire and intentions sets to generate a plan for the corresponding agent. The belief set is populated by the *Input Manager* and contains environmental data including the current location of the agent, objects and dangers around. By default, the desire of an agent is leaving the building. However, it is possible to overwrite it. For instance, an agent may give up after an unsuccessful exploration and decide to follow a particular person. Intentions contain pre-constructed abstract plans to achieve the desires. They are defined in a high level manner. For instance, when the agent detected an unblocked exit gate, it should directly go there. On the other hand, if there is a door to another room instead of an exit gate, it could navigate through the door. By using the belief

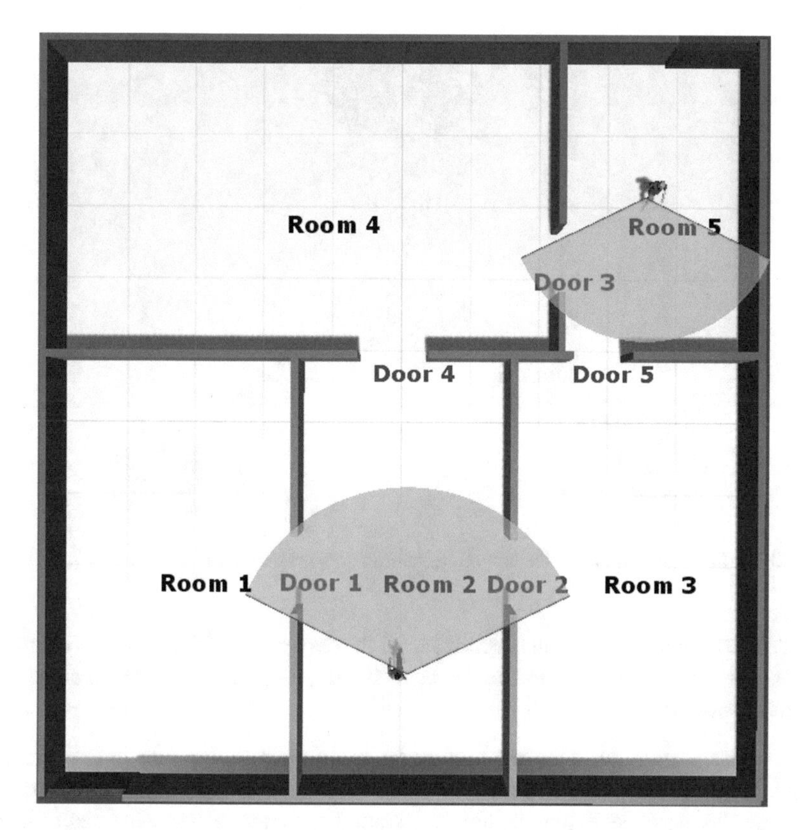

Fig. 2 Views of different evacuees

set, BDI Manager chooses the most appropriate intention that may lead to achieve the goals under the current circumstances. For this purpose, it also creates a partial map of the building using the observations. Figure 3 depicts a sample map of the bottom agent in Fig. 2 after some exploration. It shows that the agent was able to detect doors 1, 2, 4 and rooms 1, 2, 3. However, it does not know what is there after *Door 4*. Thus, it is an exploration opportunity for the agent. It may also decide to explore *Room 3* or even *Room 1* until it makes sure that there is no exit or other door(s) in this room.

(3) Action Generator *Action Generator* creates a set of appropriate actions for the current plan. For navigational actions, it uses the maps produced by the agent during exploration. For instance, if the agent decides to leave the room, action generator will check if a door is observed. If not, it will either decide to do random walk around to find an exit or follow another agent that is believed to be credible. If the agent believes that it is in danger, depending on the plans defined in the system, it will try to escape from the current location as soon as possible. Again, related actions are generated by this module, until the *BDI Manager* decides something else. Thus, this

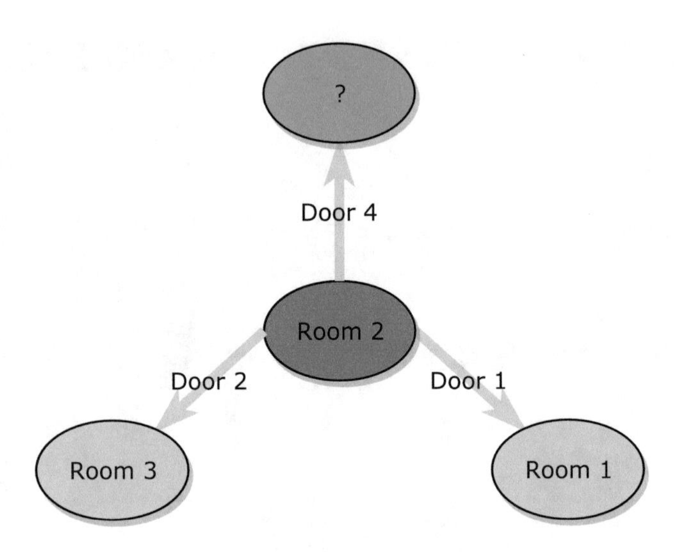

Fig. 3 Partial map example of the bottom agent after some exploration

module continuously generates actions until the current plan is achieved or changed. In case the BDI Manager updates or cancels the plan, the Action Generator starts from the beginning.

(4) Property Updater This module updates the attributes of the agent directly, depending on the current situation. Failure of attempting some actions could increase the panic level and running for some time may decrease the health level. On the other hand, any improvement in the state of the agent keeps it calm. In some cases, it may also directly affect the plan. For instance, if the agent is extremely tired, it has to take a rest in a safe place to be able to get out of the dangerous area without being hurt.

(5) Fuzzy Logic Manager This module is responsible for calculating the speed of the agent using fuzzy logic. It uses attributes of the agent in a fuzzy inference rule set after fuzzification. The resulting labels are defuzzified and crisp values are calculated.

The resulting system is highly adaptive and modular. It divides the evacuation problem into sub-problems and tries to solve them separately. It is possible to change even the major objectives. The system may continue operating under the current circumstances. Depending on the belief set, the *BDI Manager* creates a set of plans and passes it to the *Action Generator*. The action generator continuously feeds the fuzzy logic manager with actions. The *Fuzzy Logic Manager* finds the speed of the agent by using the current attribute values of the corresponding agent. It passes the final action to the agent again. The action generator generates the actions according to the current set of plans. If the BDI manager changes the plans, the action generator resets itself and starts working using the new set. This is the way to model the adaptive behavior of agents.

5 Experiments

In order to test reliability and effectiveness of our approach, we have conducted multiple experiments in a building floor with dimensions 100 m × 100 m. As depicted in Fig. 2, the floor contains five rooms and two exit gates. Eight different human models are used to make it easier to track group movements and specific individuals. Every model has its own default physical and mental attribute value ranges. The real values are assigned during initialization of the simulation. However, these values are kept the same for different experiment sets in order to eliminate having different observations as a result of attribute values. There are 140 individuals in the building. The primary objective is observing the behavior of evacuees under different configurations.

There are three different approaches we tested with multi-agent BDI framework with the fuzzy logic inference model:

- **Agents without any experience:** This setup aims to observe the behavior of agents when they have no prior knowledge about the environment. In addition, there is no cooperation among individuals. Thus, they need to find the rooms, doors and exit gates via exploration. These agents still use the BDI model, but their plan set does not contain anything related to the interaction with others.
- **Agents with partial environment knowledge:** This set of experiments aims to simulate the evacuation of individuals who have partial knowledge about the building plan. The idea is, even if it is his first time in the building, a person at least may roughly know the pathway to the gate he used to enter the building. Therefore, we assume that every individual has a plan, which is possibly not optimal, but as the crowd is not organized, it will possibly lead to some chaos.
- **Cooperative agents:** This case is for a scenario where there are experienced leaders who are knowledgeable about the evacuation and they organize the groups. These navigational agents represent educated emergency personnel. Thus, instead of moving independently, evacuees follow these leaders to egress gates. At the beginning, each evacuee starts searching for either an exit or a leader. Here it is worth mentioning that, it is possible for an agent not to find a leader and explore the area by itself as in the first test case scenario.

Each test case scenario has been simulated ten times and the number of evacuees who leave the building depending on the time graph is depicted in Fig. 4.

The results show that, in an organized set of groups that follows a trained leader, it is much faster to reach to an egress gate. They also decrease the possibility of common phenomena during evacuations, such as arching and clogging. While most of the crowd leaves the building less than 250 s, it takes more than 350 s both for exploring evacuees and partially experienced evacuees. In each set of experiments, there is a certain number of agent with low mobility. This explains the slow rate of evacuation for the latest individuals.

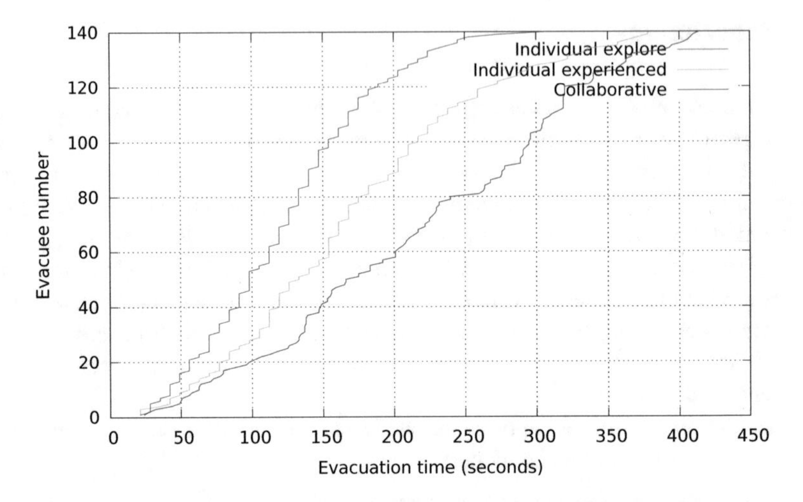

Fig. 4 Number of evacuees/evacuation time graphs for different approaches

Fig. 5 Clogging example during agent exploration

Figures 5 and 6 show snapshots of exploration agents without prior knowledge and cooperative agents respectively. Comparison between the flow in the doors for two cases clearly shows the effectiveness of organizing the groups. While explorer agents lead to clogging in the doors, the flow in the organized groups is much faster.

Fig. 6 Evacuees following navigational agents

6 Conclusions and Future Work

This work proposes a customized BDI framework for emergency evacuation simulation systems. The model is based on a multi-agent approach with a heterogeneous agent set. Depending on its abilities, each agent receives partial environment information continuously and builds an internal pool of beliefs. A pre-defined set of plans and sub-plans are filtered depending on the belief set. The most appropriate plan is selected for achieving the goal. The proposed system has been run on a test case to observe the behavior of different approaches. Our future work includes creating the plan set after some pre-exploration steps. This way, it is also possible to simulate agents with different knowledge and behavior. In addition to visual input, we are planning to propagate emotional states through neighbor agents.

References

1. R.W. Perry, A.H. Mushkatel, *Disaster Management: Warning Response and Community Relocation*. Quorum Books (Greenwood Press, Westport, 1984)
2. M. Okaya, T. Takahashi, Human relationship modeling in agent-based crowd evacuation simulation, in *International Conference on Principles and Practice of Multi-Agent Systems* (Springer, Berlin, 2011), pp. 496–507
3. H.A. Simon, *Models of Man: Social and Rational; Mathematical Essays on Rational Human Behavior in Society Setting* (Wiley, Hoboken, 1957)
4. R. Laughery, Computer simulation as tool for studying human-centered systems, in *Proceedings of the 30th Conference on Winter Simulation* (IEEE Computer Society Press, Washington, 1998), pp. 61–66

5. R.J. Dawson, R. Peppe, M. Wang, An agent-based model for risk-based flood incident management. Nat. Hazards **59**(1), 167–189 (2011)
6. Y. Bo, W. Cheng, H. Hua, L. Lijun, A multi-agent and pso based simulation for human behavior in emergency evacuation, in *2007 International Conference on Computational Intelligence and Security* (IEEE, Piscataway, 2007), pp. 296–300
7. X. Pan, C.S. Han, K. Dauber, K.H. Law, A multi-agent based framework for the simulation of human and social behaviors during emergency evacuations. AI Soc. **22**(2), 113–132 (2007)
8. J. Was, R. Lubas, Towards realistic and effective agent-based models of crowd dynamics. Neurocomputing **146**, 199–209 (2014)
9. F. Tang, A. Ren, Agent-based evacuation model incorporating fire scene and building geometry. Tsinghua Sci. Technol. **13**(5), 708–714 (2008)
10. A.S. Rao, M.P. Georgeff, Modeling rational agents within a BDI-architecture. KR **91**, 473–484 (1991)
11. M. Schut, M. Wooldridge, S. Parsons, On partially observable MDPs and BDI models, in *Foundations and Applications of Multi-Agent Systems* (Springer, Berlin, 2002), pp. 243–259
12. G. Rens, D. Moodley, A hybrid POMDP-BDI agent architecture with online stochastic planning and plan caching. Cogn. Syst. Res. **43**, 1–20 (2017)
13. R. Nair, M. Tambe, Hybrid BDI-POMDP framework for multiagent teaming. J. Artif. Intell. Res. **23**, 367–420 (2005)
14. M.P. Georgeff, F.F. Ingrand, *Decision-Making in an Embedded Reasoning System* (Citeseer, 1989)
15. M. d'Inverno, M. Luck, M. Georgeff, D. Kinny, M. Wooldridge, The dMARS architecture: a specification of the distributed multi-agent reasoning system. Auton. Agent. Multi-Agent Syst. **9**(1–2), 5–53 (2004)
16. X. Zhao, Y. Son, BDI-based human decision-making model in automated manufacturing systems. *International Journal of Modeling and Simulation* **28**(3), 347–356 (2007)
17. S. Lee, Y.-J. Son, Integrated human decision making model under belief-desire-intention framework for crowd simulation, in *Winter Simulation Conference, 2008. WSC 2008* (IEEE, Piscataway, 2008), pp. 886–894
18. D. Helbing, A mathematical model for the behavior of pedestrians. Syst. Res. Behav. Sci. **36**(4), 298–310 (1991)
19. H. Liu, B. Liu, H. Zhang, L. Li, X. Qin, G. Zhang, Crowd evacuation simulation approach based on navigation knowledge and two-layer control mechanism. Inf. Sci. **436**, 247–267 (2018)
20. R.G. Reynolds, An introduction to cultural algorithms, in *Proceedings of the Third Annual Conference on Evolutionary Programming* (World Scientific, Singapore, 1994), pp. 131–139
21. C. Adam, P. Taillandier, J. Dugdale, B. Gaudou, BDI vs FSM agents in social simulations for raising awareness in disasters: a case study in Melbourne bushfires. Int. J. Inf. Syst. Crisis Res. Manag. **9**(1), 27–44 (2017)
22. M. Valette, B. Gaudou, D. Longin, P. Taillandier, Modeling a real-case situation of egress using BDI agents with emotions and social skills, in *PRIMA 2018: Principles and Practice of Multi-Agent Systems* (Springer, Cham, 2018), pp. 3–18

Entering Their World: Using Social Media to Support Students in Modern Times

Corinne A. Green, Emily McMillan, Lachlan Munn, Caitlin Sole, and Michelle J. Eady

1 Introduction

Modern technology-rich environments provide a variety of tools with various types of capabilities that can support student success at the tertiary level. While university-supported learning platforms such as Moodle typically support this academic purpose, social networking sites such as Facebook can also be used within university studies to support student success.

One cohort of students and their academic mentor at the University of Wollongong (UOW) in New South Wales, Australia were connected together through a Facebook group. The aim of this Facebook group was to provide support to a group of students who were required to complete 1 year of a Bachelor of Social Sciences: Education for Change (BSSE4C) degree before transitioning into a Bachelor of Primary Education (BPrimEd) degree after successfully completing one full year of university study. Using free access social media software rather than a prescribed licensed program for this purpose provided a platform on which the cohort could access support from their peers and university staff, and generate a community of learners. This approach aligned with the UOW Faculty of Social Sciences Strategic Plan 2017–2021 [1] by providing "future-oriented learning experiences that meet the needs of diverse cohorts" (p. 15) and "connecting with and supporting students through all phases of the student life cycle" (p. 15).

This chapter details a 'Students as Partners in Research' project that investigated how the BSSE4C Facebook group was used by its members to support and encourage the cohort through their first year of university studies, and support their

C. A. Green (✉) · E. McMillan · L. Munn · C. Sole · M. J. Eady
University of Wollongong, Wollongong, NSW, Australia
e-mail: corinneg@uow.edu.au

M. Kaya et al. (eds.), *Putting Social Media and Networking Data in Practice for Education, Planning, Prediction and Recommendation*, Lecture Notes in Social Networks, https://doi.org/10.1007/978-3-030-33698-1_2

journey into a BPrimEd degree. It has been co-authored by some of the students and university academics involved, and reflects our collective work on this project.

2 Literature Review

The current generation of students in tertiary education have been labelled "digital natives" by some researchers for their natural affinity for and high frequency use of technology [2, 3]. While this term has been questioned [4, 5], there is undoubtedly a high use of technology within tertiary education and wider society in the modern age [6]. Rapid communication technology, such as mobile phones and social networking sites, and web resources were amongst the most frequent types of technology used by university students [7]. A large portion of the student population regularly access social media daily, with the most popular and frequently accessed sites being Facebook, Twitter, Instagram, and YouTube [8]. Bicen and Cavus [9] found that most students were spending at least 4 h on Facebook a day, using the site as a social tool through messaging and their news feed to keep in contact and stay up to date with their friends, through a variety of devices.

Social networking sites such as Facebook are often regarded as having a sole purpose as a communication tool amongst friends [10, 11]. However, there is evidence of students using these sites as an academic tool, to facilitate communication between peers, and to share and review work [3]. In fact, Gallardo-Echenique et al. [12] found that students were much more likely to communicate with their peers through social networking sites and other publically available ICT applications than university-supported applications.

In light of the frequent use of technology within the student population, Gallardo-Echenique et al. [13] advocate fruitful discussions of the needs of digital learners, how staff can respond their needs, and how technologies can be designed to be responsive to the needs of the digital learner. The notion of needing a completely new learning environment for this new generation is refuted by Romero et al. [14], given the lack of evidence showing support in the difference in how this current and previous generations are taught [4, 6]. Instead, they recommend an adaptation of information to make the content more accessible [14]. In a similar vein, Gallardo-Echenique et al. [3, 13] suggest that universities should make use of social networks to take advantage of the communicative opportunities and positive student attitude toward these networks.

Facebook has frequently been used in studies to understand the potential usefulness of the site as part of academia [15–18]. Broadly speaking, students have held positive attitudes towards using Facebook as part of their degree [19]. Ratneswary and Rasiah [18] found Facebook to be a good communicative tool and platform on which students could share their ideas and knowledge regarding assessments, which also helped to build a sense of belonging among the student cohort, breaking down potential barriers among the students as well as lecturers [15]. Using Facebook as the platform for this communication enabled immediate feedback [16, 17], important for students seeking information about assessments [20, 21]. The familiarity that students have with Facebook, as evidenced by its daily use by the majority of the student population [9], increased its usefulness as an academic tool [16, 20].

However, the use of the platform in this way raises concerns for some regarding privacy as well as the use of a social space as a teaching space [15, 21, 22]. Furthermore, a study conducted by Jacobsen and Forste [23] demonstrated a negative association between the use of electronic media (such as social networking sites, video games, or texting) and students' grades within university subjects. It is worth noting that students and academics have differing perspectives regarding such technology within the university setting, both with regards to the legitimacy of its use and the impact of it on students' learning [24, 25]. Ariel and Elishar-Malka [24] advocate a recognition of the value social media and technology can contribute to students' learning, "rather than aspiring to make the 'nuisance' go away" (p. 10).

Previous research into the use of Facebook within university degrees has frequently investigated the use of 'Facebook pages', which can be viewed by the public with anyone able to 'like' the page and receive updates [19, 22, 23]. Very few studies explore the use of 'Facebook groups', where a community can be formed and restrictions can be placed on who can participate [20, 27]. Furthermore, much of the research explores using Facebook within a specific subject, with class materials and information specific to the content of lectures or tutorials posted [28, 29]. The student response to this has been mixed, given that these class materials were already accessible on university-supported learning platforms such as Moodle or Blackboard [16, 29]. It has been shown that using Facebook in this way could cause it to become a distraction and hinder study, rather than support student achievement [17, 20, 27]. In response, Cooke [30], Donlan [22] and Petrovic et al. [17] recommend that Facebook be used as a supplementary tool alongside university-supported learning platforms, providing administrative updates in regard to the course and spreading and sharing information.

3 Context

As a result of a recommendation within the Great Teaching, Inspired Learning report [31], ratified by the NSW Education Standards Authority [32], new restrictions applied to entrants into NSW undergraduate teaching programs from 2016. These restrictions meant that prospective teacher education students needed to have achieved marks above 80 (i.e. Band 5) in three subjects in their High School Certificate (HSC), with one of these in English. Alternatively, the successful completion of 1 year (48 credit points) of university would suffice to grant them entry into a Bachelor of Education degree.

The students involved in this study had not met these criteria, and therefore were unable to enrol in the Bachelor of Primary Education (BPrimEd) degree at UOW in 2016. They were given the option to enrol instead in a course called Bachelor of Social Science: Education for Change (BSSE4C). This undergraduate course would enable students to study eight 6 credit point subjects across two semesters in 2016. Provided they successfully completed these subjects, the students could then transfer to the BPrimEd and still complete their degree in the 4 year time frame.

Michelle (ME) took on the role of the 'go to person' for the BSSE4C student cohort in an effort to support and encourage the students as they completed their first year of studies, and aid them in their transition into the BPrimEd. As ME was not involved in face-to-face teaching of the cohort within the BSSE4C degree, she did not have the access to the usual university-supported means of communication. In an effort to keep the students motivated, focused and in touch with the School of Education, the students were asked at an initial face to face meeting if they would like to have a Facebook group. The cohort agreed and the BSSE4C Facebook Group was created. Students were added to the group by request, with ME as administrator ensuring that each request was from a BSSE4C student. All students enrolled in BSSE4C joined the Facebook group, with the end goal of transferring to the BPrimEd degree always at the forefront of its use. In 2017, 106 students from BSSE4C successfully transferred into the BPrimEd degree.

3.1 Students as Partners in Research

With the apparent success of the Facebook group in supporting students to successfully complete their first year of university studies and transfer into the BPrimEd degree, a research project was proposed. This investigation into the Facebook group's use, and the interactions between group members it facilitated, would add to the emerging understandings of how Facebook and other social networking sites can be used effectively within tertiary studies.

The student cohort was invited to be significantly involved in the project through a 'Students as Partners in Research' approach. Six students agreed to co-author a paper about the experience with two members of university staff. This collaborative approach to conducting the research enabled insights from the student perspective to be incorporated into the research process [33]. It also presented a great opportunity for the undergraduate students to learn about the research and publication process from start to finish [34]. Three of these students, along with the two university academics, are the co-authors of this paper, with the other three choosing to withdraw to focus on their university studies.

4 Methodology

This study was a mixed-method investigation into the use of a Facebook group with a specific cohort of students studying at the University of Wollongong in 2016. Data was collected through screenshots of posts made within the Facebook group between February 2016 and February 2017, and a survey of the cohort in June 2017. The survey was distributed to students via their university email and included questions about participants' previous use of Facebook for university studies and the help they received (or not) through the BSSE4C Facebook group in their first year of study. Informed consent for the use of this data was granted by 102 of 126 students in the cohort.

To maintain participant confidentiality, responses captured within the Facebook screenshots and the survey have been de-identified and coded as follows. In the code 'SP.43', 'SP' denotes survey participant, followed by the participant number. Codes such as 'FB.JF' are used for Facebook group contributions, where 'FB' denotes Facebook, followed by the participants' initials. The posts and comments of those who did not give consent were blacked out prior to data analysis.

Iterative coding was used with both the Facebook data and survey responses. Sentiment analyses were not pursued, given the limitations of automated and text-mining approaches [35, 36]. As Clarke and Mehmet [36] acknowledge, sentiment and emotion are not easily determined through algorithms that do not have a functional interpretation of language. Instead, the Facebook data and survey responses were analysed by the research team using a technique common within qualitative research [37]. For each dataset, participants' responses were coded, and emerging themes generated. As a result of the ongoing process of data collection and theme formation, a code book emerged and was followed by the research team throughout all stages of the data analysis. Once all of the themes where identified, a process of reducing the themes and amalgamating similar areas to common themes for reporting took place [37].

5 Results

The survey data and data pulled from the Facebook group itself have been combined to provide a comprehensive picture of the value of this Facebook group as students transitioned through their first year at university, and then into the BPrimEd degree.

Of the 126 of students in the BSSE4C cohort in 2016, 102 gave consent for their posts and comments within the BSSE4C Facebook group to be used in this research project. Of these, 59 students also completed a survey about their use of the Facebook group and its value.

The quantitative data is presented first, followed by qualitative data according to themes.

5.1 Quantitative Data

The majority of survey respondents (91.5%) had a Facebook account prior to commencing their university studies. Just over one fifth of survey respondents (22%) reported using Facebook for academic purposes other than the BSSE4C Facebook group, such as to access learning resources, connect with other students, and within group work projects. SP.3 outlined the value of Facebook for these purposes, stating that "We were able to have a group discussion and keep each other informed on where we were up to without having to wait for email replies". With regards to the BSSE4C Facebook group in particular, an overwhelming number of survey respondents (95%) declared that it helped them through their first year of study.

Across the 13 months from the inception of the Facebook group until the end of data collection (February 2016–February 2017), a total of 674 posts were made on the Facebook group, of which 70% incorporated a question. March 2016 had the largest number of posts—25% of the total number of posts—with another significant spike in posts in February 2017 (see Fig. 1). These months represent the beginning of an academic year.

A variety of interaction methods were employed by the Facebook group members. These include asking questions within posts or comments (1107 instances in total), using emojis (755 instances), and posting photos (92 instances), links (36 instances), or pictures such as screenshots or memes (32 instances). Group

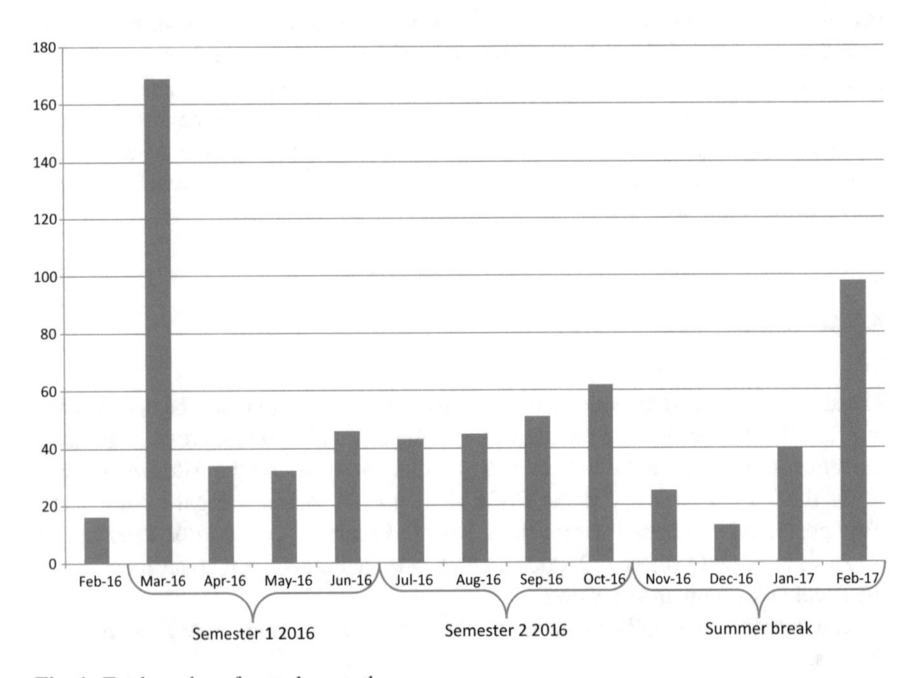

Fig. 1 Total number of posts by month

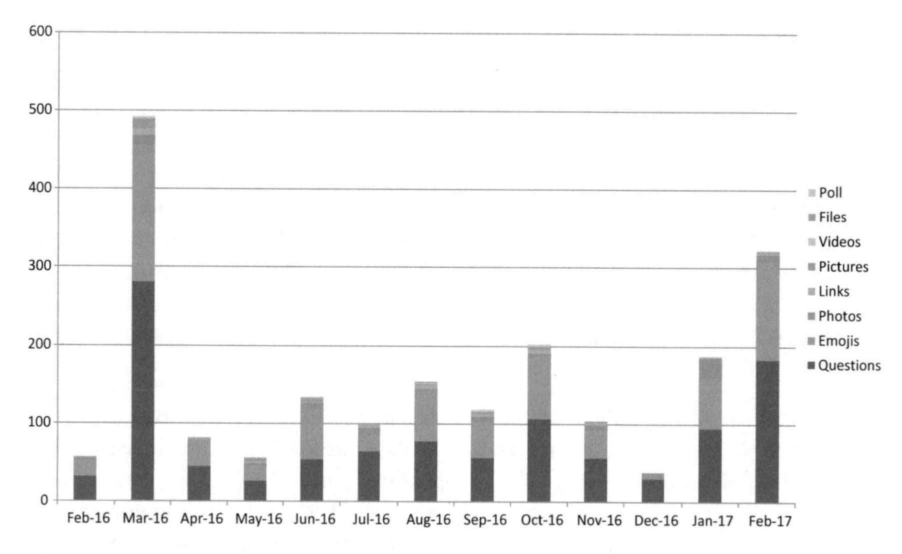

Fig. 2 Interactions on BSSE4C Facebook group by month

members also uploaded videos (10 instances) or files (6 instances), and polled group members (4 instances) (see Fig. 2). Only 5% of posts (35 posts) had no signs of engagement from group members via comments or any of the aforementioned means of interaction.

5.2 Qualitative Data

Thematic analysis of the Facebook group data and the survey responses reveals three main purposes of the Facebook group for those involved. First and foremost, it provided a space for students to ask questions about their university experience, for instance regarding their assessment tasks and the expectations of their subject tutors. Linked to this is the involvement of the university academic, who provided a supporting role (rather than a face to face teaching position) for the BSSE4C students over the data collection period. Finally, the Facebook group facilitated a community for the students where they could find peer support and know they were not alone.

Overall, the responses to the survey indicate that students had a positive attitude towards the Facebook group as they transitioned into university studies, and as they transferred from the BSSE4C degree into the BPrimEd degree. This sentiment is illustrated in responses such as "I truly do believe that our Facebook group has been a great asset to my success in this degree" (SP.22) and that the group "has really helped me make the most of my degree" (SP.56). The students indicated that this type of Facebook group should be available in other courses, with one stating "I am personally disappointed that the course I'm in now does not have something like this. I feel it is valuable for future students" (SP.3).

Questions About the University Experience According to the BSSE4C students, a significant value of the Facebook group was found in asking questions of their peers related to their university studies. It was frequently noted in the survey responses that the Facebook group "provided a platform … where we were able to ask questions and share information with each other" (SP.28) because "everyone had questions, everyone could answer and it was really helpful to have a question answered within seconds!" (SP.58). The speed with which students' questions could be answered was a positive for students, as "a lot of the time students reply to Facebook quicker than tutors or subject coordinators do through emails" (SP.8).

Students found that others "tended to ask questions that I was also thinking" (SP.22) and that, typically, "if anyone has asked a tutor/coordinator a question they let everyone else know" (SP.8). One respondent particularly noted that "having people upload a response from a particular lecturer in regards to an assignment … would benefit everyone and reduce the amount of emails that the lecturer would have to respond to" (SP.52). This was evident throughout the Facebook group, such as in October when it became clear that the deadline for an assessment task had changed abruptly. FB.EG commented, "I emailed [the subject coordinator] about it, as we're in the same boat, to see if the late penalty marks will still apply. I mentioned that other students have the issue of not being able to submit it in time" and promised to "send word when she replies".

Students posed questions on a range of academic issues, including subject content, assessment submission processes, and study techniques. However, one student expressed disappointment in their peers' use of the Facebook group, declaring that "a lot of the questions asked were really stupid and could be found in the subject outline … I got the feeling that people were using it to be too lazy to look for the answers themselves" (SP.42). Around the time of tutorial enrolments each semester a number of posts requesting tutorial swaps would be largely ignored, with no interaction evident.

Involvement of the University Academic Having an academic (ME) in the Facebook group in a supporting role provided an opportunity for someone who is knowledgeable of university processes to speak into the discussions that students were having. ME was called upon to clarify certain situations and to "provide helpful and correct information" (SP.31). This occurred early in the year when students were confused about their TurnItIn similarity reports—"I got 0% too! ME is this bad??" (FB.BM)—and more frequently towards the end of the year as misinformation regarding the BSSE4C cohort's transfer into the BPrimEd degree abounded. In one instance, FB.TL calls on ME to "please clear this up", to which ME responded (in under an hour, outside of standard work hours), "Ok deep breaths everyone" before proceeding to explain the issue.

The students were deeply grateful for this level of interaction from the academic, making comments such as "we'd be lost without you" (FB.GH), and "Thanks

ME, we all appreciate your effort" (FB.CP). Even when frustrated by university processes, FB.GH commented that she was "not blaming ME for the record, she's a 10/10 legend". The actions of ME through the Facebook group helped build rapport with the students, who declared that this "has improved my learning with her ... in 2017" (SP.56) and that "I believe [this relationship] improves our chances of success ... I'm so happy to have had [ME] as our cheerleader in the first year and subject coordinator in the second year" (SP.58).

From this position of trusted authority figure, ME was able to point out the positive strides that students were making towards their chosen degrees and careers. In response to an online interaction between several students, ME commented "I LOVE how you all work together! Collaboration is a massive part of being a teacher! Way to go!!" Similarly, when one student helped another with a maths question, ME commented "LA = teacher material!!! AJ = professional lifelong learner!!! Love it you two! You've made my day!!" ME encouraged the students in their academic work, sharing with them when she heard "lots of positive talk [amongst other subject coordinators and academics] about the BSSE4C group", and impressing upon the cohort that they could succeed (see Fig. 3).

Survey responses indicated that the focus of the Facebook group shifted somewhat once the academic removed herself from the group when she began teaching the cohort in 2017. SP.10 stated that "once the group no longer had someone who was an authority figure in it, comments, posts and the general usage of the site started to become unprofessional. Photos and videos of uni related information were quickly replaced with posts of people out drinking and partying."

The presence of the academic in the group increased its usefulness for students, who stated it enabled them to "effectively [raise] an issue with staff" (SP.15)" and "provided additional support. With this support I felt like we always had someone there looking out for us" (SP.56).

Community of Students The BSSE4C Facebook group facilitated the cohort to develop into a community of learners who could support, encourage, and empathise with one another throughout their degree. Students explicitly mentioned how the Facebook group had "allowed me to meet some people" (SP.22) in their cohort, "allowed for stronger relationships to be built within our course" (SP.28), and "helped me to be better acquainted with my cohort and be comfortable talking to them outside and in classes" (SP.21). According to SP.39, "the Facebook page still plays a big role in our sense of community as a cohort today".

While there were a few social posts, such as when FB.KT invited her peers to the pub to celebrate the completion of their first university semester, the content of the group was primarily focused on academic matters. For SP.7, this was the main benefit of the group: "I think the [group] was most effective in the sense that as a student, we could talk about subjects and uni in a less professional sense". The members of the Facebook group were connected as students, and this was reflected in their use of the group.

There were multiple posts across the year that reflected the opportunity the group gave the cohort to share their frustrations and confusion and, in so doing, find

Fig. 3 FB.ME post from June, 2016

support and advice from their peers. Questions related to the mechanics of university study, such as the nature of TurnItIn similarity reports, were met by comments like "Mine did that too" (FB.NP). Students were able to vent their frustrations about assessment tasks and marking procedures, leading others to empathise with comments such as "I completely agree with you! So frustrating" (FB.TCM) and "[It's] not just you" (FB.EG). FB.RI's post asking for confirmation that "because this exam is only 30% you don't have to get a minimum of 45%?" received responses that encouraged him to "Aim higher" (FB.KT), and that also showed he was not alone in his feelings: "I feel the same" (FB.NP). As SP.53 summarised, students found it "nice to know people were going through the same emotions as me at times, and [I] loved that everyone stuck together and helped each other when someone didn't understand something or required extra help in a particular topic".

Students willingly gave answers to their peers' questions and offered advice throughout the year through the platform provided by the Facebook group. At the beginning of the year, FB.EC gave some insight into her own study techniques (such as completing subject readings and re-writing lecture notes). She encouraged others, saying "I already know my style. It's good to get started but don't stress, it takes time to figure it out" (FB.EC). When the Semester 1 exam period was approaching, FB.CP recognised the emotional toll it might take, posting "I'm sure there will be much frustration but this will all be worth it, just charge through and do the best you can". As the paperwork for the cohort's transition into the BPrimEd degree became available, students shared the process with one another to make it as smooth as possible for all involved. For SP.59, the advice and encouragement offered by their peers through the Facebook group "really helped, knowing people going through the same thing as you were there to help".

6 Discussion

Through the analysis of data from the BSSE4C Facebook group and survey responses it is evident that the Facebook group provided support to students through their first year of university studies and as they transitioned into the BPrimEd degree. The Facebook group provided a forum where students could go to ask questions about their studies, connect with a university academic, and build a community of learners. There was a high level of interaction between group members throughout the 13 month period of data collection, especially at the beginning of the academic year. 95% of posts were engaged with, through questions, comments, emojis, memes, and other forms of interaction.

The vast majority of participants in this study were already familiar with Facebook prior to their enrolment in BSSE4C (91.5%), even for academic purposes (22%). Given this, and the fact that social networking sites function well as a communicative tool, it was an ideal platform to use. The student response to the use of Facebook in the university setting was generally positive, echoing the findings of research conducted by Al-Mashaqbeh [19], Ellefsen [16] and Ratneswary and Rasiah [18]. The results also showed similarity to that of Alarabiat and Al-Mohammad's [15] study as a sense of belonging was developed, with students commenting that using the Facebook group "allowed for stronger relationships to be built within our course" (SP.28).

The use of a Facebook group (involving a degree cohort), rather than a Facebook page (created for a particular subject), resulted in a supplementary tool that could be used alongside existing learning platforms. This approach differs from much of the previous research into using Facebook within university studies [19, 22, 26] and was recommended by Cooke [30], Donlan [22] and Petrovic et al. [17].

The results of this study show that using Facebook as a university communication tool is a good way to create a supportive and interactive space where students can come together and receive help and encouragement. It allowed students to collectively find answers to their university-related queries, with input provided by their peers and a university academic. This echoes the findings of VanDoorn and

Eklund [21], where immediate answers to questions assisted students in their studies and enabled them to spend more productive time working on assessment tasks.

The usefulness of the group was enhanced by the presence of the academic, who frequently provided encouragement as well as clarification when needed. As acknowledged by Akcaoglu and Bowman [20] and Bowman and Akcaoglu [28], the continued contributions of this individual to the online discussion throughout the year played an important role in fostering conversations and monitoring the use of the group throughout the year. The impact of this role in the BSSE4C Facebook group was highlighted by its reported shift in focus in the absence of such an authority figure.

The Facebook group was enjoyed by the vast majority of students, with collegial relationships and a sense of community developing within the cohort as a result of its use. It provided a space for students to air their grievances, share common experiences, and encourage one another through their first year of university study. Similar findings were established by Bowman and Akcaoglu [28], Donlan [22] and Ratneswary and Rasiah [18].

7 Conclusion

The findings of this study indicate the potential for using Facebook groups within university studies, particularly to provide support to a particular degree cohort during a time when extra encouragement is needed. Capitalising on the power of social networking sites such as Facebook in this way can provide a modern platform for communication between and amongst students and academics, with the aim of supporting student success.

Further research into the impact of Facebook groups on student achievement and other student outcomes, as well as with additional cohorts in other disciplinary areas, would be of value to tertiary educators seeking to support students. The inclusion of a university academic in this group, and the incidental impact they appeared to have on the group's focus and use is particularly interesting. Comparative studies investigating the nature of Facebook groups with and without academic input, or the impact of an academic who was required by their superiors to interact with the Facebook group, as opposed to an academic who volunteered to interact with the Facebook group, may bring further understanding to this area.

Finally, conducting this research as a 'Students as Partners in Research' project has been a rewarding experience for all of the authors. The collaboration between university academics and students broadened our understandings of this project and the use of the Facebook group with the BSSE4C cohort. We were able to share our diverse perspectives as students, university academic involved with the Facebook group, and university academic who became involved after the Facebook group was handed over to the student cohort. This chapter represents a year of working together on this project, and is a culmination of our collective work and writing.

References

1. University of Wollongong, *Faculty of Social Sciences Strategic Plan 2017–2021* (University of Wollongong, Wollongong, 2017)
2. C. Brown, L. Czerniewicz, Debunking the 'digital native': beyond digital apartheid, towards digital democracy. J. Comput. Assist. Learn. **26**(5), 357–369 (2010). https://doi.org/10.1111/j.1365-2729.2010.00369.x
3. E.E. Gallardo-Echenique, L. Marqués Molías, M. Bullen, Students in higher education: social and academic uses of digital technology. Int. J. Educ. Technol. High. Educ. **12**(1), 25–37 (2015). https://doi.org/10.7238/rusc.v12i1.2078
4. S. Bennett, K. Maton, Beyond the 'digital natives' debate: towards a more nuanced understanding of students' technology experiences. J. Comput. Assist. Learn. **26**(5), 321–331 (2010). https://doi.org/10.1111/j.1365-2729.2010.00360.x
5. E.J. Helsper, R. Eynon, Digital natives: where is the evidence? Br. Educ. Res. J. **36**(3), 503–520 (2010). https://doi.org/10.1080/01411920902989227
6. G. Kennedy, T. Judd, B. Dalgarno, J. Waycott, Beyond natives and immigrants: exploring types of net generation students. J. Comput. Assist. Learn. **26**(5), 332–343 (2010). https://doi.org/10.1111/j.1365-2729.2010.00371.x
7. P. Thompson, The digital natives as learners: technology use patterns and approaches to learning. Comput. Educ. **65**, 12–33 (2013). https://doi.org/10.1016/j.compedu.2012.12.022
8. A. Al-Bahrani, D. Patel, B. Sheridan, Engaging students using social media: THE students' perspective. Int. Rev. Econ. Educ. **19**, 36–50 (2015)
9. H. Bicen, N. Cavus, Social network sites usage habits of undergraduate students: case study of Facebook. Proc. Soc. Behav. Sci. **28**, 943–947 (2011). https://doi.org/10.1016/j.sbspro.2011.11.174
10. Y. Amichai-Hamburger, G. Vinitzky, Social network use and personality. Comput. Hum. Behav. **26**(6), 1289–1295 (2010). https://doi.org/10.1016/j.chb.2010.03.018
11. C. Ross, E.S. Orr, M. Sisic, J.M. Arseneault, M.G. Simmering, R.R. Orr, Personality and motivations associated with Facebook use. Comput. Hum. Behav. **25**(2), 578–586 (2009). https://doi.org/10.1016/j.chb.2008.12.024
12. E.E. Gallardo-Echenique, M. Bullen, L. Marqués-Molías, Student communication and study habits of first-year university students in the digital era. Can. J. Learn. Technol. **42**(1), 1–21 (2016). https://doi.org/10.21432/T2D047
13. E.E. Gallardo-Echenique, L. Marqués-Molías, M. Bullen, J.W. Strijbos, Let's talk about digital learners in the digital era. Int. Rev. Res. Open Dist. Learn. **16**(3), 156–187 (2015)
14. M. Romero, M. Guitert, A. Sangrà, M. Bullen, Do UOC students fit in the net generation profile? An approach to their habits in ICT use. Int. Rev. Res. Open Dist. Learn. **14**(3), 158–181 (2013)
15. A. Alarabiat, S. Al-Mohammad, The potential for Facebook application in undergraduate learning: a study of Jordanian students. Interdiscip. J. Inf. Knowl. Manag. **10**, 81–103 (2015)
16. L. Ellefsen, An investigation into perceptions of Facebook use in higher education. Int. J. High. Educ. **5**(1), 160–172 (2016)
17. N. Petrovic, V. Jeremic, M. Cirovic, Z. Radojicic, N. Milenkovic, Facebook versus moodle in practice. Am. J. Dist. Educ. **28**(2), 117–125 (2014)
18. R. Ratneswary, V. Rasiah, Transformative higher education teaching and learning: using social media in a team-based learning environment. Proc. Soc. Behav. Sci. **123**, 369–379 (2014)
19. I. Al-Mashaqbeh, Facebook applications to promote academic achievement: Student's attitudes towards the use of Facebook as a learning tool. Int. J. Modern Educ. Comput. Sci. **7**(11), 60–66 (2015)
20. M. Akcaoglu, N. Bowman, Using instructor-led Facebook groups to enhance students' perception of course content. Comput. Hum. Behav. **65**, 582–590 (2016)

21. G. VanDoorn, A. Eklund, Face to Facebook: social media and the learning and teaching potential of symmetrical, synchronous communication. J. Univ. Teach. Learn. Pract. **10**(1), 6 (2013)
22. L. Donlan, Exploring the views of students on the use of Facebook in university teaching and learning. J. Furth. High. Educ. **38**(4), 572–588 (2014)
23. W.C. Jacobsen, R. Forste, The wired generation: academic and social outcomes of electronic media use among university students. Cyberpsychol. Behav. Soc. Netw. **14**(5), 275 (2011). https://doi.org/10.1089/cyber.2010.0135
24. Y. Ariel, V. Elishar-Malka, Learning in the smartphone era: viewpoints and perceptions on both sides of the lectern. Educ. Inf. Technol. **24**(1), 1–12 (2019). https://doi.org/10.1007/s10639-019-09871-w
25. L.R. Elliot-Dorans, To ban or not to ban? The effect of permissive versus restrictive laptop policies on student outcomes and teaching evaluations. Comput. Educ. **126**, 183–200 (2018). https://doi.org/10.1016/j.compedu.2018.07.008
26. G. Jenkins, K. Lyons, R. Bridgstock, L. Carr, Like our page: using Facebook to support first year students in their transition to higher education (A practice report). Int. J. First Year High. Educ. **3**(2), 65–72 (2012)
27. H. Chiroma, N. Shuib, A. Abubakar, A. Zeki, A. Gital, T. Herawan, J. Abawajy, Advances in teaching and learning on Facebook in higher institutions. IEEE Access **5**, 480–500 (2017)
28. N.D. Bowman, M. Akcaoglu, "I see smart people!": using Facebook to supplement cognitive and affective learning in the university mass lecture. Internet High. Educ. **23**, 1–8 (2014). https://doi.org/10.1016/j.iheduc.2014.05.003
29. E. Dolan, E. Hancock, A. Wareing, An evaluation of online learning to teach practical competencies in undergraduate health science students. Internet High. Educ. **24**, 21–25 (2015). https://doi.org/10.1016/j.iheduc.2014.09.003
30. S. Cooke, Social teaching: student perspectives on the inclusion of social media in higher education. Educ. Inf. Technol. **22**(1), 255–269 (2017)
31. NSW Department of Education and Communities, *Great Teaching, Inspired Learning: A Blueprint for Action* (Board of Studies, Teaching and Educational Standards, the NSW Department of Education and Communities, the Catholic Education Commission NSW and the Association for Independent Schools NSW, Sydney, 2013)
32. NSW Education Standards Authority, Increased Academic Standards for Studying Teaching. Future Teachers (2017), http://educationstandards.nsw.edu.au. Mar 2018
33. E. Nelson, P. Bishop, Students as action research partners: a New Zealand example: authors examine how one teacher in New Zealand engaged in collaborative action research with her students. Middle Sch. J. **2**, 19 (2013)
34. P. Fieldsend-Danks, The dialogues project: students as partners in developing research-engaged learning. Art Des. Commun. High. Educ. **15**(1), 89–102 (2016). https://doi.org/10.1386/adch.15.1.89_1
35. A.I. Canhoto, Y. Padmanabhan, 'We (don't) know how you feel': a comparative study of automated vs. manual analysis of social media conversations. J. Mark. Manage. **31**(9–10), 1147–1157 (2015). https://doi.org/10.1080/0267257X.2015.1047466
36. R.J. Clarke, M.I. Mehmet, *Semantics of Textual Sentiment: Recovering Attitudes in Facebook Marketing Messages*, Collaboration Laboratory Working Paper 6 (2016)
37. A. Bryman, *Social Research Methods*, 5th edn. (Oxford University Press, Oxford, 2016)

Utilizing Multilingual Social Media Analysis for Food Venue Recommendation

Panote Siriaraya, Yuanyuan Wang, Yukiko Kawai, Yusuke Nakaoka, and Toyokazu Akiyama

1 Introduction

In the past decade, the rapid growth of the telecommunication infrastructure has led to a significant proportion of the population to adopt online technology to augment various aspects of their daily lives [1]. Such users communicate with their friends and family members through online platforms, share and receive news through social media sites such as Facebook and carry out informed purchasing and consumption behaviors through aggregated review sites. The advances and proliferation of mobile technology have also enabled social information to be available in a more ubiquitous manner. For example, users are able to receive information about how others use, rate and perceive the various establishments within their geographical vicinity through location-based services which utilize social data.

The vast amount of social information which has emerged from the rapid adoption of such platforms has proven to be extremely beneficial for both industrial practice and scientific research in several ways. First, such data has enabled researchers to develop various approaches which allow them to analyze and predict user behavior. For example, by analyzing the movement trajectories of different

P. Siriaraya (✉)
Kyoto Institute of Technology, Kyoto, Japan
e-mail: spanote@kit.ac.jp

Y. Wang
Yamaguchi Daigaku, Ube, Japan
e-mail: y.wang@yamaguchi-u.ac.jp

Y. Kawai · Y. Nakaoka · T. Akiyama
Kyoto Sangyo Daigaku, Kyoto, Japan
e-mail: kawai@cc.kyoto-su.ac.jp; g1444936@cc.kyoto-su.ac.jp; akiyama@cc.kyoto-su.ac.jp

M. Kaya et al. (eds.), *Putting Social Media and Networking Data in Practice for Education, Planning, Prediction and Recommendation*, Lecture Notes in Social Networks, https://doi.org/10.1007/978-3-030-33698-1_3

users as they "check in" to various facilities, researchers have been able to obtain a better understanding of the characteristics of a particular urban area [2]. In domains such as health and safety, researchers have even shown how information from social media platforms such as Twitter could be useful in helping detect natural disaster events such as earthquakes [3] or to monitor prescription medicine abuse [4]. In the meanwhile, in business and commerce, consumers have become increasingly reliant on social metrics (such as aggregated user ratings and user comments about a particular service or product) to make informed purchasing behaviors [5].

While most of these examples show how spatial, time, interaction and even emotion based-information from social media platforms could be used to analyze user behavior or to predict user interest, few studies have investigated how contextual information related to cultural aspects (such as similarities in their nationalities) from social media platforms could be used in the prediction of user preferences, especially in systems designed to recommend products and services to users (such as [6, 7]). A key problem which is generally encountered in such research is the sparsity of social media information in less populated areas for a specific population group (i.e. there are few French geo-tagged tweets in locations such as Berlin or Rome). Thus, the applicability of this approach would be somewhat limited to areas with a dense social media footprint.

In this paper, we report the results of a study carried out to develop a system which takes into account the linguistic properties of information (e.g., the language of users) and uses data from social media platforms to suggest food venues to users. We leveraged the communication language used on social media as a proxy of culture traits and utilize such data in the recommendation of food venues for users. More specifically, the venue recommendation system was developed to propose food venues for users in a specific city by utilizing information from the micro-blogging service Twitter as well as the location-based social media platform Foursquare. To help cope with situations of sparse data, our system was designed to recommend food venues in locations where few geo-tagged tweets are available. This was done by calculating the similarity between different language users in their genre preferences by utilizing an approach similar to collaborative filtering. Users of the same language will receive a similar recommendation of venues. Overall, this paper is an extended edition of our previous conference paper published in [8] where we extend upon our prior work by examining the use of tweet sentiment analysis to enhance the recommendation process. In addition, a more thorough analysis of our proposed approach was carried out where French speakers who had visited or lived in five different European cities were asked to rank the top ten venues generated using different types of algorithms to help compare the performance of our algorithm. While our prior evaluation was calculated based on the re-ranking of a pre-generated top ten venue list, the algorithms in this study were evaluated based on top ten venues generated individually for each algorithm. This required an expansion of the data collection to gather information about the rating scores and surrounding tweets for all the venues in our data set as well as a new round of data collection with 55 more users.

Overall, the main research objectives in this study could be summarized as follows:

Obj1: Develop a system which utilizes contextual properties, that is the language of users as a proxy of culture trait to recommend food venues which can cope with situations of sparse data.

Obj2: Investigate how objective user information such as venue rating and subjective user information such as tweet sentiment could be used to enhance the recommendation of the food venues.

Obj3: Examine how users would perceive the venues recommended by such a system through user evaluated crowd sourcing.

2 Related Work

Prior studies have shown how information from social media could be used to detect various events and phenomena which occur in the physical or virtual space and how such information could be used to help enhance our understanding of human behavior. For example, these studies show how the spatial and temporal properties of geo-tagged tweets could be used to help detect trends [9, 10] or track various events [11]. One study even proposed a system in which an open-domain calendar of significant events could be created by analyzing content from Twitter [12]. Other researchers have proposed different ways social media could be used as a type of early warning system, namely, to detect unexpected events such as earthquakes [3], disease outbreaks [13] and accidents [14]. Large-scale systems which utilize data from social network platforms have also been proposed as a way to provide real-time insight into the emotional status of entire communities [15].

In human mobility research, data from location-based social media platforms have been used to allow us to better understand human mobility behavior in the real world. Recent studies tend to address the issue of extracting and identifying mobility patterns from Twitter, by analyzing the content as well as the spatial and temporal aspects of individual records to infer the general movement patterns of people in a geographical location [16–18]. Such information has, for example, enabled researchers to better understand how demographic groups affect urban mobility [19]. Further studies show how large-scale analysis of social media data enables us to better understand global trends in human mobility, such as the effect of season on mobility patterns and the characteristics of international travel in different nations [20]. The practical uses of such information also allow systems to be developed which could recommend the safest route for travelers [21] or help identify the trade area of a specific store (e.g., a restaurant or a shop) [22].

As Internet access has become increasingly available to the global population, the number of non-English speaking Internet users have also increased to become a significant proportion of the online population. Thus, multicultural and multilingual studies among social media users have emerged to become a niche research topic in social network analysis research. Such a topic is becoming increasingly important for social media research, as a lot of information is spread across different languages and countries and research which exclude such information could miss out on a significant proportion of valuable information. While prior studies have been carried out in this field, most focus on the technical aspects, such as devising a model to determine the language of choice of a specific user [23], or creating classification models to determine the number of languages inside a particular tweet [24, 25]. However, there have been few studies which utilize this multi-linguistic information in combination with spatial information or user experience data (such as sentiment or service ratings) to predict user interest or behavior.

Also, sentiment analysis has become a popular tool for data analysts, especially those who deal with social media data. It has become quite common to look for public opinions and reviews of services, events, and products on social media using computational approaches. Excellent surveys of the field have established that rich online resources have greatly expanded opportunities for opinion mining and sentiment analysis [26, 27]. Liu [28] introduces an efficient method, at the state of the art, for doing sentiment analysis and subjectivity in English. Jijkoun et al. [29] focus on generating topic specific sentiment lexicons. However, most of the existing sentiment analysis methods such as sentiment lexicons were designed for universal English without the focus on particular geographic areas. It is however, crucial to develop new technology to be able to adapt sentiment analysis to a wide number of other cultures and areas [30–32].

In this paper, we aim to develop a food venue recommendation system which utilizes cultural traits (by using communication language as a proxy) as well as objective information such as user venue visits and subjective information such as user ratings and tweet sentiment in the nearby vicinity. Such data is obtained by analyzing information from two social media platforms, Twitter and Foursquare. While various approaches for recommending products or services exist, most rely on explicit user interactions and behaviors (such as users actively rating products, making purchases, consuming services) [33, 34]. Other systems utilize contextual information about the characteristics of the users themselves (such as age and gender) from their client database to further enhance recommendation accuracy [35]. However, these systems tend to focus only on information available from within their own service platform (user ratings or text comments from their database) and do not capitalize on the wealth of mobility, sentiment and perception data available in modern social media systems.

For recommendation systems which do focus on social media data, most do not distinguish between different language users (such studies tend to focus on a single language and are tested using data from English speaking cities). For instance, one study developed a recommender system based on check-in and tips data from Foursquare in London and New York [36]. Another examined the use of property data from the venues and user comments obtained from Foursquare to recommend points of interests in the area of California [37]. More recent studies used image data from Instagram to augment check-in data from Foursquare to recommend venues for users in Chicago and New York [38]. Overall, these systems generally do not take into account the multi-linguistic properties of information from social media systems in their recommendation. Our work is unique in that it incorporates spatio-temporal information with linguistic, emotional and user perception information from two large-scale social media platforms in the analysis of user food preferences and uses such information to recommend food venues for users.

3 Food Venue Recommendation Based on Geo-Tagged Tweets

In this section, we discuss the method used to suggest food venues to users in our study. An overview of the process used in the proposed system is as follows:

Step 1: For each city, identify the food venues which have been visited by each language user by analyzing information from Geo-tagged tweets

Step 2: For every language user, calculate the evaluation score for each food genre based on their food venue visits

Step 3: Calculate the degree of similarity between each language user in their food genre preferences using Eq. (3).

Step 4: For each language, calculate the evaluation score for every food venue by using two types of equations (see steps 5 and 6) based on a certain threshold.

Step 5: If the number of venues visited in a city for a specific language user is below the threshold value α (i.e., there are not enough visitors of a particular language for meaningful analysis), use Eq. (2) to extract food genres which have a high chance of being favored by users and then suggest food venues which correspond to those extracted genres.

Step 6: If the number of venues identified (for a specific language at a city) is above the threshold α, use Eq. (1) to suggest food venues.

A visual overview of our proposed method is provided in Fig. 1.

3.1 Extracting Venues from Geo-Tagged Tweets

The first step of our approach is to determine which venues in a city had been visited by a specific language user. Information about the location, time and language is extracted from the geo-tagged tweets collected in the area as well as information about the language of the tweet. In analyzing the language of the tweets, we examine two types of language information which are available on Twitter: The mother tongue language parameter on the user profile and the language used in the tweet itself. A tweet is considered to be related to a particular language when either the mother tongue parameter or tweet language matches that language.

To extract the food venues which had been visited by each language user, we focused on 'I'm at' type tweets (a special type of tweets which users post to

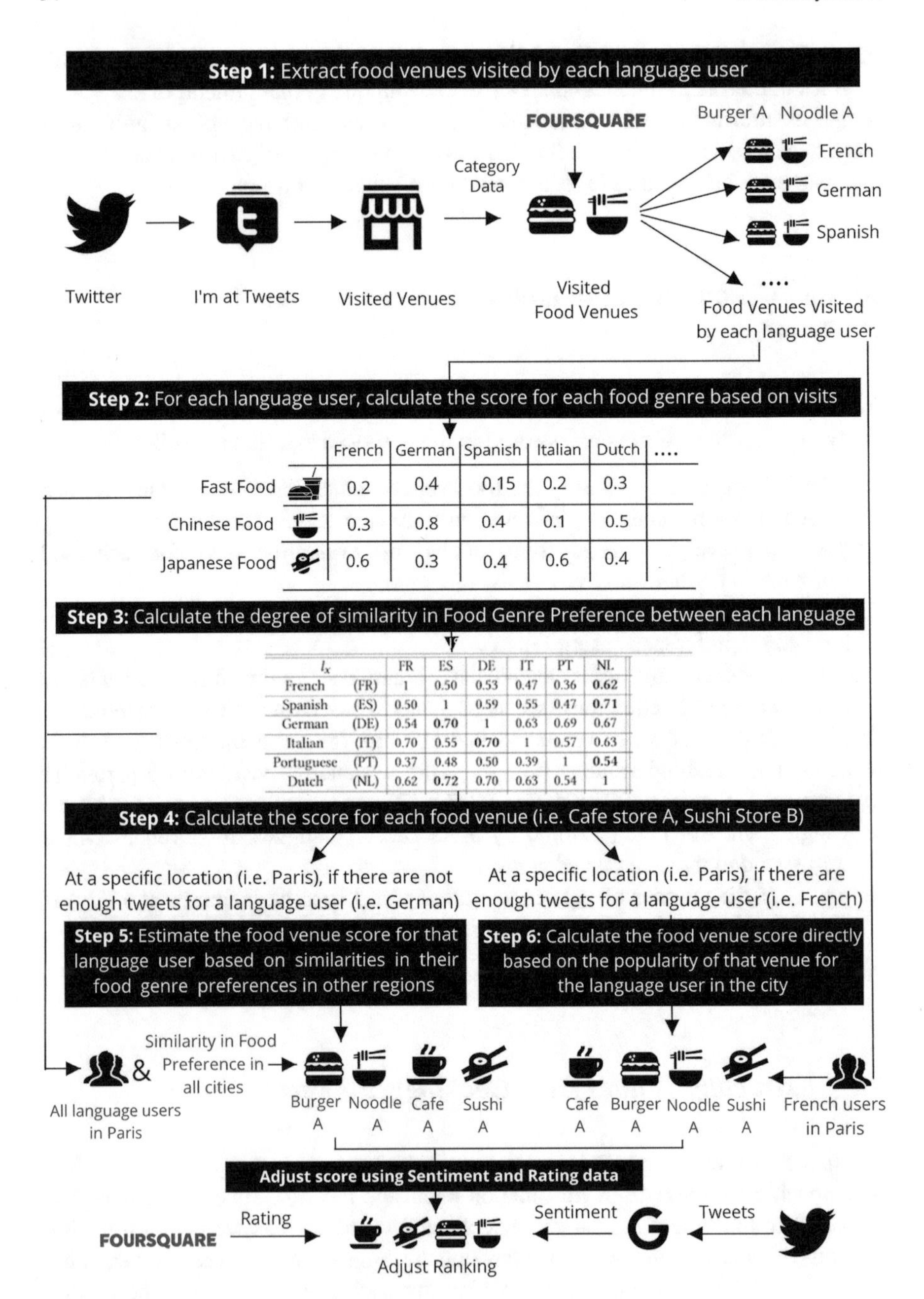

l_x		FR	ES	DE	IT	PT	NL
French	(FR)	1	0.50	0.53	0.47	0.36	**0.62**
Spanish	(ES)	0.50	1	0.59	0.55	0.47	**0.71**
German	(DE)	0.54	**0.70**	1	0.63	0.69	0.67
Italian	(IT)	0.70	0.55	**0.70**	1	0.57	0.63
Portuguese	(PT)	0.37	0.48	0.50	0.39	1	**0.54**
Dutch	(NL)	0.62	**0.72**	0.70	0.63	0.54	1

Fig. 1 An overview of the method proposed in the paper

denote that they are at a specific location). Data about the geographical coordinates (latitude-longitude) are collected from these tweets. In addition, we extract data about the category (or genre) of food venues contained in the 'I'm at' tweets by utilizing the Swarm API from Foursquare[1] to search for the venue category based on the venue name detected from the tweet and the geographical coordinate. The genre data used in this study is the second level category value in the 'food' category in the foursquare data. For example, the category of 'food' contains second level categories such as "Chinese food" and "cafe".

3.2 Calculating the Venue Score Based on Popularity

If a sufficient number of tweets is detected in a city (**Step 6**) for a specific language user (i.e., # tweets > threshold α), the following Eq. (1) is used to calculate the food venue score. Overall, the popularity of a venue for a given language user is used to represent the venue score for that language user (using a TF-IDF based approach).

$$S_{\{i,p\}} = \frac{|T \in p : l_y \in T : i \in I_t|}{|T \in p : l_y \in T|} \cdot \log \frac{|L|}{|l \in L_i|} \tag{1}$$

where $|T \in p : l_y \in T : i \in I_t|$ refers to the number of tweets sent from the city p using the language l_y in venue i, $|T \in p : l_y \in T|$ refers to the total number of venues of tweets sent from the location p using the language l_y, $|L|$ refers to the total number of languages L and $|l \in L_i|$ refers to the number of languages in venue i.

3.3 Calculating the Venue Scores for Locations with Sparse Tweets for a Given Language User

In most cases however, there is not a sufficient amount of tweets by a specific language user for meaningful analysis of food preference for the region (i.e., # tweets < threshold α). In such cases, we estimate the food venue preferences of that language user based on similarities in their food genre preferences with other language users from all other regions in our dataset (**Step 3**). For example, in a region where there are few French tweets, we infer the food genre score for French language users in that region by examining the similarity in food genre preferences between French language users and other language users (Spanish, Dutch, etc.) in other regions. This food genre score is then used to determine which food venues French language users would likely be interested in. More specifically, the degree of

[1]https://developer.foursquare.com/.

similarity in food genre preference *sim(x,y)* between two language users in specific countries (country x and country y) is calculated using the following Eq. (2):

$$sim(x, y) = \frac{\sum^{J}(GF_{\{x,j\}} - \overline{GF_{\{x,j\}}})(GF_{\{y,j\}} - \overline{GF_{\{y,j\}}})}{\sqrt{\sum(GF_{\{x,j\}} - \overline{GF_{\{x,j\}}})^2 \sum(GF_{\{y,j\}} - \overline{GF_{\{y,j\}}})^2}} \tag{2}$$

where $GF_{\{x,j\}}$ denotes the evaluation score of a specific genre j in the language l_x and $GF_{\{y,j\}}$ denotes the evaluation score of the other language l_y. The evaluation score of the food genres in the city (i.e. $GF_{\{x,j\}}$) would be calculated using Eq. (3), based on the popularity of a specific genre for that language user in a country (**Step 2**):

$$GenreFrequency_{\{x,j\}} = \frac{\#\text{Genre } j \text{ in the language } l_x}{\text{The combined \#all genres in } l_x} \tag{3}$$

The food venues which have the highest frequency (top n food venues) for each recommended genre (i.e., those with the highest simulated genre evaluation score) are used to provide food venue recommendations for the user (**Step 5**). For example, in Berlin, if the food Genres "Cafe", "Italian Restaurant" and "Sea Food Restaurant" are calculated to have the highest similarity based genre evaluation score for French people, then when recommending the top 6 food venues, our system would recommend Cafe Extrablatt, March Bistro (the top 2 visited Venues in the Cafe Genre), Bella Itali, Capri Italiano (the top 2 visited Venues in the Italian Restaurant Genre) and Austernbar and Oyster Bar KaDeWe (the top 2 visited Venues in the Sea Food Restaurant Genre).

4 Adjusting Food Venue Recommendation Through Explicit Rating and Sentiment Information

Whilst popularity could be a good indicator as to the value of a food venue for a specific user, we would also like to incorporate the quality of the store in our recommendation approach as well. Venues in the Fast Food genre, for example, are specifically designed to cater to a large number of dinners and could thus artificially inflate the number of visitors detected by geo-tagged tweets (for example, when compared to food venues such as high-end restaurants which aim to cater only to a few dinners in a given day). Therefore, we incorporated more subjective measures such as the ratings from users as well as the sentiment of nearby tweets to adjust the food venue score rankings in our system.

The rating score of each venue was obtained using the Foursquare Places API.[2] These rating scores were obtained for all the food venues identified in Sects. 3.2 and 3.3 (**Step 5** and **Step 6**). As there could be multiple branches of a venue in a given city (Starbucks in La Rambla, Barcelona and Starbucks in Maremagnum, Barcelona etc), the overall rating score of a food venue was calculated by aggregating the rating score of each detected food venues in our system and normalizing the value with the overall number of raters (see Eq. (4)). We decided to use the aggregated score for a specific venue type due to the sparsity of rating data on specific branches and to match our proposed recommendation method in Sect. 3 (which also considered different branches as the same venue). Min-max normalization is further carried out as the rating scores obtained from Foursquare tend to cluster around the 4.0–9.0 range. The adjusted rating venue score is then calculated based on the popularity score obtained previously and the overall rating score. This is shown in the following Eq. (5):

$$RS_{v,l} = \frac{\#Raters_b * AveRS_{b,v}}{total\#Raters_v} \tag{4}$$

$$AdjRatingVS_{\{v,l\}} = \log\left(\frac{VS_{v,l} - VSmin_l}{VSmax_l - VSmin_l}\right) * \frac{RS_{v,l} - RSmin_l}{RSmax_l - RSmin_l} \tag{5}$$

where $RS_{v,l}$ refers to the rating score for venue v for the language user l, $AveRS_{b,v}$ refer to the average rating score for a particular branch of the venue which was obtained from the Foursquare API, $\#Raters_b$ refers to the number of raters which rated that branch and $total\#Raters_v$ refers to the total number of raters for all branches of that venue. Finally, $VS_{v,l}$ refers to the Venue Score for venue v for the language user l (which was calculated based on **Step 5** or **Step 6**) and $AdjRatingVS_{\{v,l\}}$ refers to the adjusted score which is used to rank the venues to recommend to users. Note, if no rating data is available on Foursquare, then the particular venue is excluded from the ranking.

In addition to the explicit rating scores provided by the users in each venue, the sentiment of the tweets posted nearby the venue was also used as a proxy measure for the quality of the store. In location-based recommender systems, prior studies have argued how tweet sentiment could be used to generalize positive or negative experiences at a particular location [21, 36]. More positive tweets in the area would indicate that users are enjoying themselves in the particular venue and could thus be considered as an indirect measure as to the quality of the venue. Therefore, we collected tweets within 20 m in the vicinity of each food store venue and analyzed the sentiment of the tweet content. This distance is similar to those used in other studies which match data from social media to spatial locations [39] to help account

[2]https://developer.foursquare.com/places-api.

for the error margin from GPS devices (which tend to be approximately 5–10 m in open or closed building locations [40]).

The Google Cloud Natural Language API[3] was used to carry out sentiment analysis, due to the diversity of the languages in the tweets. Pre-processing was carried out by removing HTML links and converting all @ tags in the tweet text into proper nouns. The processed text of each tweet surrounding the food venues were then sent to the Cloud Natural Language Service. The service would then return a sentiment score for each tweet. Scores between −1 to 0 represented negative sentiment and 0 to 1 represented positive sentiment. For example, the tweets "Best Moccaccino" and "The best oysters in Berlin!" received a score of 0.7 and 0.6 respectively which represented a positive experience and tweets such as "The stool here is too small" and "Disappointing Italian dinner at <name of the store>" received a score of −0.6 and −0.9 which represented a generally negative experience. Similar to the rating scores, the overall sentiment score of a particular venue was calculated by averaging the sentiment scores of the tweets obtained nearby all the branches of the venues in a particular city which were detected by our system. Finally, the adjusted sentiment venue score was calculated using the following Eq. (6):

$$AdjSentimentVS_{\{v,l\}} = \log\left(\frac{VS_{v,l} - VSmin_l}{VSmax_l - VSmin_l}\right) * AveSS_{v,l} \tag{6}$$

where $AveSS_{v,l}$ refers to the average sentiment score for a given venue and $VS_{v,l}$ refers to the Venue Score for venue v for the language user l. In short, we calculated the sentiment score from all the tweets that had been matched to a particular venue and averaged the result. This average sentiment score was then used as a weight to adjust the previously calculated venue score. It should be noted that min-max normalization was not carried out on average sentiment score as the range of Sentiment scores of the tweets did not show a skewed distribution. Also, if no tweets are detected in the vicinity, then the particular venue is excluded from the ranking.

5 System Implementation

To examine the feasibility and performance of our proposed approach, geo-tagged tweets from several European cities (due to the high diversity of native languages) were collected, during April 1st, 2016 to April 30th, 2017 (representing a 13 month period). In particular, we focused on tweets from the following seven cities in Europe: (1) London, (2) Rome, (3) Paris, (4) Barcelona, (5) Berlin, (6) Lisbon, and (7) Amsterdam in the following six languages: (1) Italian, (2) French, (3) Spanish, (4) German, (5) Portuguese, and (6) Dutch. We chose specifically to exclude English

[3]https://cloud.google.com/natural-language/.

Table 1 The number of geo-tagged tweets collected and analyzed and the total number of unique food venues identified

Language	#Tweets	"I'm at" tweets (%)	#Total venues (%)
All	25,993,771	1,231,980 (4.7%)	342,992 (1.3%)
Italian	2,251,204	98,488 (3.6%)	36,940 (1.6%)
French	2,430,737	36,163 (1.4%)	29,851 (1.2%)
Spanish	4,801,999	40,367 (0.8%)	34,813 (0.7%)
German	2,041,920	216,242 (8.6%)	55,414 (2.7%)
Portuguese	881,874	24,585 (2.8%)	22,359 (2.5%)
Dutch	1,671,522	257,383 (15.4%)	269,413 (16.1%)
Total	14,079,256	673,228 (4.8%)	448,790 (3.2%)

Table 2 The number of unique food venues identified at seven European cities (the underlined value represents the number of unique food venues in the city which is home to that language)

Language	London	Rome	Paris	Barcelona	Berlin	Lisbon	Amsterdam
Italian	2914	<u>6203</u>	369	1706	81	39	153
French	1568	363	<u>16,445</u>	797	5	157	209
Spanish	3624	3419	868	<u>20,614</u>	117	240	464
German	1454	367	211	820	<u>873</u>	44	276
Portuguese	634	115	479	373	131	<u>2127</u>	313
Dutch	197	67	368	261	68	101	<u>3165</u>
Total	10,391	10,534	18,750	24,571	1275	2708	4580

as (1) it was the common language (i.e. the language used in communication between people who do not share a similar first language) that was most used throughout the world and (2) was a language used by a large number of culturally diverse nationalities. This makes it difficult to meaningfully distinguish whether the tweets from this language truly reflected the characteristics of a specific user type. As such, the venues in our system were recommended using Italian, French, Spanish, German, Portuguese and Dutch tweets.

Overall, approximately 26 million tweets and 0.34 million different venues were identified. Tables 1 and 2 shows the number of geo-tagged tweets collected in each city and language as well as the number of venues identified from the 'I'm at tweets'. Although, the data was usually dense in cases where the tweet language was the same as the language used by the native population of the city (Italian in Rome, French in Paris, etc.), in cases where the language was different, there was sparse data in a majority of the cities (Italian in Paris, Portuguese in Rome, etc.). In Berlin, for example, only five venue tweets were identified for French language users. A total of 155 different genres were detected using our approach.

Table 3 shows an example of the venues recommended by our system for French language users based on (1) language independent popularity (i.e., ranked according to the number of I'm at tweets obtained for all language users in a particular city)

Table 3 The top ten venues recommended for five different cities for French language users based on the different ranking approaches

City	Adjustment	Top ten food venues
Amsterdam	All languages	Mcdonalds, Starbucks, Hard Rock Cafe Amsterdam, The doner company, Burger Bar, Cafe de Barones, Wok to Talk, Vondelpark3, FEBO
	None adjusted	Cafe de Barones, Restaurant Cafe in de Waag, Starbucks, De Koffieschenkerij, Cafe 't Papeneiland, Cafe 't Smalle, Mcdonalds, Burger King, Vondelpark3, Amsterdam Hard rock Cafe
	Rating adjusted	De Koffieschenkerij, Cafe 't Smalle, Starbucks, Cafe 't Jaren, Cafe 't Papeneiland, Cafe 't van Zuylen, Vondelpark3, Caffe ll Momento, Cafe de Barones, Cafe van Kerkwijk
	Sentiment adjusted	Caffe Ristretto, De Koffieschenkerij, Temple Bar, De Silveren Spiegel, Supperclub Cruise, Douwe Egberts, McDonalds, Bar Spek, Bridges Cocktail Bar at Sofitel Amsterdam, Back to Black
Berlin	All languages	Dunkin' Donuts, goodies Berlin, eismanufaktur Berlin, MatrixClub Berlin, Cafe Extrablatt, March Bistro, Cafe & Bar Celona Bella italia, TeeGschwendner, Austernbar
	None adjusted	Cafe Extrablatt, March Bistro, Bella Italia, Capri Italiano, Austernbar, Oyster Bar KaDeWe, Kilkenny Irish Pub, Blarney Irish Pub, Kartoffelacker, Zum Hackepeter
	Rating adjusted	Austernbar, Cafe & Bar Celona, Lir Irish Pub, Bella Italia, Nuova Mirabella, Kartoffelacker, Celtic Cottage—The Irish Pub, Irish Pub, Cafe Extrablatt, Finnegan's Irish Pub
	Sentiment adjusted	Casa Italiana, Blarney Irish Pub, Galleria illy, Sarah Wiener, Austernbar, Dachgarten, Irish Pub, Hummer & Langusten, Cafe im Zeughaus, Hopfingerbrau im Palais
Barcelona	All languages	Pans & Company, Granier, La Tagliatella, McDonalds, El Fornet d'en Rossend, 100 Montaditos, Camp Nou Dinner Terrasse, Tapas24 Camp Nou, Restaurante Park Gaell, La Paradeta Passeig de Gracia
	None adjusted	100 Montaditos, McDonald's, La Paradeta Passeig de Gracia, Marco Aldany, Central Cafe, Hard Rock Cafe Barcelona, Gran Cafe, Hidden Cafe, El Tastet de la Mar, El Merendero de la Mari
	Rating adjusted	La Paradeta Passeig de Gracia, Hard Rock Cafe Barcelona, 100 Montaditos, Hidden Cafe, Restaurante Barceloneta, La Muscleria, Brunch & Cake, Picnic, Cerveseria Ciutat Comtal, Alsur Cafe

(continued)

Table 3 (continued)

City	Adjustment	Top ten food venues
	Sentiment adjusted	El Masteler De La Barceloneta, La Paradeta Passeig de Gracia, 100 Montaditos, McDonald's, Picnic, La Muscleria, Restaurante Chino China Town, Restaurante Namaste, Hidden Cafe, Cafe Godot
Rome	All languages	McDonald's, Numbs, Don Chisciotte, L a Piazzetta, Canova Piazza del Popolo, Bar Gelateria Fontana Di Trevi SRL, Sant'Eustachio Il Caffe, Vecchia Roma, La Bottega del Caffe, La Biga Wine Food
	None adjusted	Vecchia Roma, La Bottega del Caffe, Canova - Piazza del Popolo, Sant'Eustachio Il Caffe, La Piazzetta, Ristorante La Scuderia, il Pastarito Ristorante Pizzeria, Ristorante Al Boschetto, McDonald's, Ristorante Ciao
	Rating adjusted	Sant'Eustachio il Caffe, Ai Tre Scalini, Antico Caffe Greco, Canova-Piazza del Popolo, La Boccaccia, Angelina a Trevi, il Piccolo Buco, Ristorante La Pentolaccia, Ai Balestrari, Ristorante Alessio,
	Sentiment adjusted	Ristorante Pizzeria Santa Croce, Ai Balestrari In Prati, Sant'Eustachio II Caffe, Ponte e Parione, Ristorante Il Fungo, Taverna Urbana, Ristorante Cucina Pepe, Gran Caffe Laura Laura, Il Corallo, Caffe del Teatro
Lisbon	All languages	Quasi Pronti, Time Out Market Lisbon, Estefania Portugal, Rui dos Pregos, Portela Cafes, Fabrica dos Pastais de Belam, Nosolo Italia Praa do Comercio, Restaurante & Bar Terreiro do Paso, Blue Lounge, Casa Brasileira
	None adjusted	Taberna da Rua das Flores, Bella Lisa Rossio, A Brasileira, Cafetaria Museu do Teatro, Restaurante & Bar Terreiro do Pao, Blend Bairro Alto, Lisboa Bar, Lisbona Bar, Quasi Pronti, Nosolo Italia Praa do Comercio
	Rating adjusted	Taberna da Rua das Flores, Hard Rock Cafe (Ins Meneses), Taberna Moderna, Taberna Sal Grosso, Fabrica Coffee Roasters, Taberna Santa Marta, Pois, Cafe, Cafe da Garagem, O Prego da Peixaria, Choupana Caffe
	Sentiment adjusted	O Pardieiro, Le Petit Cafe Restaurante, Restaurante Aviz, Tasca do Jaime, Lisbona Bar, Morgadinho de Alfama, Pastelaria Tim Tim, Solar dos Presuntos, Restaurante S.Miguel de Alfama, Procapio,

(2) the none-adjusted venue scores (calculated in **Step 5** and **Step 6**), (3) the rating adjusted venue scores and (4) the sentiment adjusted venue scores (see Sect. 4).

Finally, to visualize the results of our study, a map-based interactive food venue recommendation system was also developed as a web application (see Fig. 2). When using the system, users would first select a language. Afterward, markers would be shown on the map to represent the recommended food venues for users of that

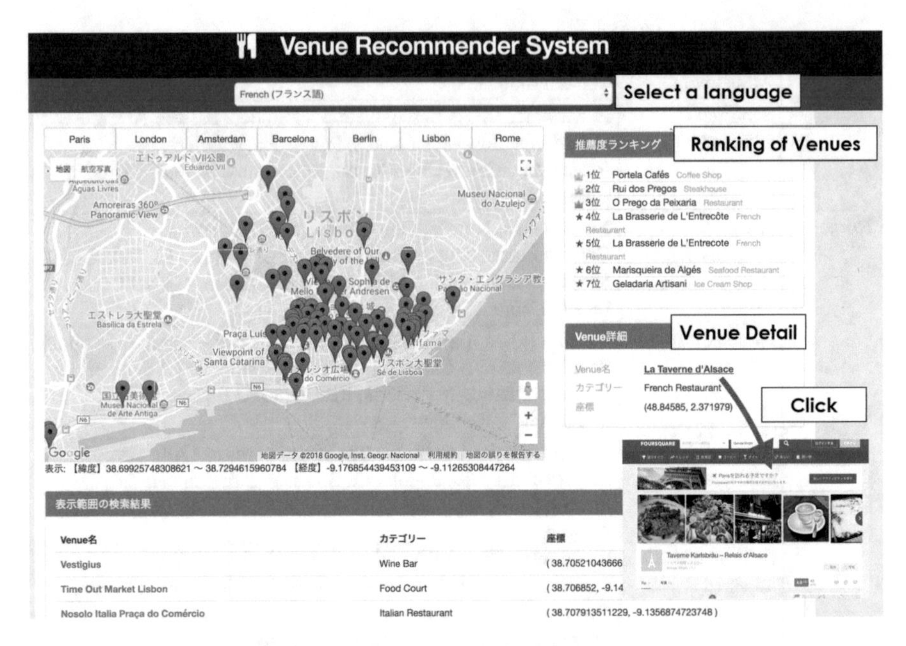

Fig. 2 A map based web app implementation of our venue recommender system

language in a particular city. On the right side of the map, the rankings of the venues would be shown. Users would be able to read the information about the venue by moving their mouse over the marker.

6 User Evaluation Study

To examine the performance of the proposed recommendation method, people who had visited or lived in the cities in which twitter data was collected were asked to rate their preferences of the restaurants recommended using our approach. The crowd sourcing Internet marketplace platform mechanical turk (MTurk)[4] was used to recruit participants. This platform has shown to be an efficient means of gathering good quality data for research in a number of fields [41, 42]. Participants were asked to fill in a short questionnaire where they were asked to rate how likely they were to visit the different food venues recommended using the different ranking approaches on a 7-point Likert scale (see Table 3 for the venues which participants were asked to rate). Figure 3 shows an example of the questionnaire used in the study. Those who completed the questionnaire were given \$0.25 to \$0.35 as compensation based on the length of the questionnaire (which is comparable in accordance to

[4]https://www.mturk.com/mturk/welcome.

Veuillez evaluer la probabilité pour vous de visiter les restaurants (-3 = très improbable, 3=très probable) *

Si vous ne connaissez pas le restaurant, veuillez cliquer sur le lien à côté du nom du restaurant

	-3	-2	-1	0	1	2	3
Back to Black (http://goo.gl/XPB6x9)	○	○	○	○	○	○	○
Bar Spek (http://goo.gl/dh8F52)	○	○	○	○	○	○	○
Bridges Cocktail Bar at Sofitel Amsterdam (http://goo.gl/EacLzN)	○	○	○	○	○	○	○
Café de Jaren (http://goo.gl/TpHrtY)	○	○	○	○	○	○	○
Café van Kerkwijk (http://goo.gl/cmcUYR)	○	○	○	○	○	○	○
Café Van Zuylen (http://goo.gl/wDiZCw)	○	○	○	○	○	○	○
Caffè Il Momento (http://goo.gl/2qEaCs)	○	○	○	○	○	○	○
Caffe Ristretto (http://goo.gl/Ys9pfj)	○	○	○	○	○	○	○
De Silveren Spiegel (http://goo.gl/4JB7Bj)	○	○	○	○	○	○	○
Douwe Egberts Kiosk (http://goo.gl/zxVeWL)	○	○	○	○	○	○	○
FEBO (http://goo.gl/7Cp3Ce)	○	○	○	○	○	○	○
Restaurant Café In de Waag (http://goo.gl/A2x8ZZ)	○	○	○	○	○	○	○
Supperclub Cruise (http://goo.gl/XHwbSo)	○	○	○	○	○	○	○

Fig. 3 The questionnaire used in the evaluation for French language users in Amsterdam

MTurk's mandatory compensation policies for the expected amount of time spent). Overall, we chose to focus on French language users and on five European cities (Amsterdam, Berlin, Barcelona, Lisbon, and Rome) due to (1) the expected sample size available on Mechanical Turk (Questionnaires for other language users (Dutch, German etc.) which were published on M-Turk but received a low participation rate) and (2) the varied distribution in the number of venue tweets identified in the different cities. We did not include Paris as we wanted to focus our investigation on French speakers in cities where French is not the native language and London as we excluded English language tweets from our analysis.

Overall, two rounds of data collection were carried out, in which (round 1) participants were asked to rate the top ten venues recommended in the none readjustment condition and (round 2) rated the top ten venues recommended in the readjustment and baseline conditions (totaling 30 venues). In the first round, a total of 65 participants responded to the questionnaire (13 participants from Berlin, 11 from Lisbon, 12 from Amsterdam, 16 from Rome, 13 from Barcelona). In the second round, a total of 55 participants responded to the questionnaire (8 participants from Berlin, 15 from Lisbon, 10 from Amsterdam, 13 from Rome and 9 from Barcelona). Four different conditions were examined, three based on our proposed algorithm and one based on a baseline approach (see Table 3 for more details about the venues which were recommended in the different conditions).

Baseline (All languages): The first condition represented the baseline condition where recommendation was based mainly on the popularity of the stores identified from the tweets (without considering the language). The food venues which had the most tweets regardless of language had the highest score.

No re-adjustment: The second condition considered the language of the tweets when calculating the score of the venues. The threshold α or the number of tweet venues required for sufficient analysis using the approach outlined in **Sect. 3.2** was set to 500 in this experiment (During our earlier study [8], we tested different parameters and this value showed the best result). As such, when recommending venues to French speakers, Eq. (3) (in Sect. 3.2) was utilized for the cities of Amsterdam, Berlin, Rome, and Lisbon and Eq. (2) (in Sect. 3.3) was utilized for the city of Barcelona.

Rating based re-adjustment: The third condition utilized the explicit user ratings to re-adjust the venue recommendation (as described in Sect. 4)

Sentiment based re-adjustment: The fourth condition utilized sentiment scores to re-adjust the venue recommendation (as described in Sect. 4)

Table 4 shows the number of tweets which were detected in the vicinity of the venues identified for French language users as well as the number of venues which contained Foursquare rating values. Both data were used in the rating and sentiment based re-adjustments. It should be noted that the low percentage of venues with Sentiment analyzed Tweets and Foursquare data for Berlin could be attributed to the generally low social media footprint of the city in comparison to other European cities. In addition, the low percentage of venues with Tweets for Amsterdam is due

Table 4 The number of tweets and Foursquare ratings available for the venues in the different European cities evaluated in our study

City	#Tweets	Venues with tweets (%)	Venues with Foursquare rating (%)
Amsterdam	2519	34.00	61.92
Berlin	655	35.60	42.86
Lisbon	5239	47.01	54.93
Rome	44,571	60.78	46.79
Barcelona	54,339	77.40	59.70

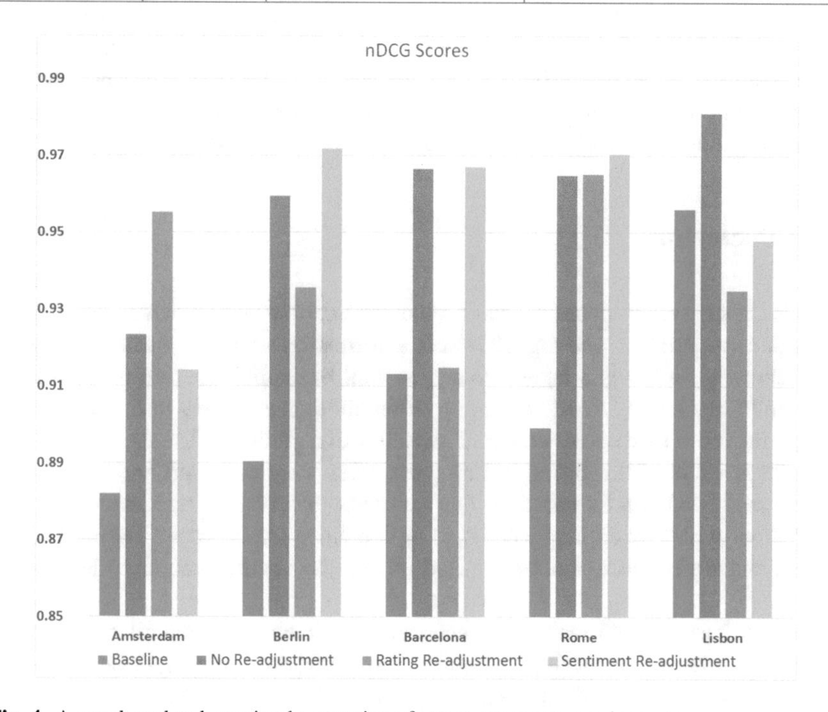

Fig. 4 A map based web app implementation of our venue recommender system

to the inability of the sentiment analysis system (discussed in Sect. 4) to analyze tweets in the Dutch language.

To evaluate the ranking quality of our proposed venue recommendation approach, the normalized Discounted Cumulative Gain (nDCG) was calculated based on the user evaluations from the French speakers for the top 10 recommended venues. The likelihood in which participants rated seeing themselves visit the various venues were used to represent their perceived relevance of the recommended venues. Figure 4 provides a summary of the results. Overall, the results showed that our proposed algorithm, which took into account the linguistic characteristics of users out-performed the baseline approach in all five cities. The nDCG score of our none adjusted algorithm was 0.923 (vs the baseline score of 0.882) for Amsterdam, 0.96 (vs the baseline score of 0.89) for Berlin, 0.97 (vs the baseline score of 0.91) for Barcelona, 0.96 (vs the baseline of 0.90) for Rome and 0.98 (vs the baseline score of 0.96) for Lisbon. Both the Rating Re-adjusted approach and the Sentiment Re-adjusted approach also outperformed the baseline approach in four out of five cities (except for Lisbon). However, incorporating explicit rating data and sentiment information in the score calculation did not result in a superior performance in the ranking quality for all the cities. The Rating based re-adjustment performed best only for the city of Amsterdam (0.955 vs 0.923 for the no re-adjustment), the Sentiment based Re-adjustment performed best for the cities of Berlin, Barcelona (0.967 vs 0.966 for the no re-adjustment) and Rome (0.97 vs 0.964 for the no re-adjustment). The Performance gains for the cities of Barcelona and Rome, however, were marginal. In addition, for the city of Lisbon, the no re-adjustment condition performed best when compared to the re-adjusted conditions (0.98 vs 0.93 for the Rating re-adjustment and 0.948 for the Sentiment re-adjustment).

7 Conclusion

In this paper, we proposed a venue recommendation system which utilizes the linguistic properties of geo-tagged tweets. Information about the location, language and sentiment of the geo-tagged tweets as well as explicit user rating data could be used to recommend food venues for different language users in different cities, even if the cities themselves contain few tweets of a particular language. Data from 26 million tweets from different European cities were collected and analyzed to recommend food venues based on our approach. Afterwards, user evaluation was carried out with French language users in five cities (Amsterdam, Berlin, Lisbon, Rome, and Barcelona) who were asked to rate the venues recommended by our system.

Four different conditions were evaluated: (1) a baseline condition where the venues were ranked according to popularity regardless of language characteristics, (2) a condition based on the approach proposed in this paper, (3) a condition utilizing explicit user ratings to re-adjust the venue ranking score and (4) a condition utilizing the sentiment score of the tweets nearby the venue to re-adjust the venue ranking score. The results showed that our proposed approach outperformed the baseline approach for all five cities for French language users. The value of using rating and sentiment scores in the rankings was less clear however, as the rating score outperformed the none-adjusted approach only for the cities of Amsterdam and Rome and the sentiment based re-adjustment out performed the none-adjusted approach only for the cities of Berlin, Barcelona and Rome.

Overall, our work provides several contributions. First, we discuss how a food venue recommender system could be developed by analyzing information from location-based social network platforms within a multilingual context. Although, existing product recommendation systems do exist, most rely on explicit user interaction information already existing in a closed platform to make recommendations for users (i.e., [33, 34, 43]). Our system however utilizes publicly available information from existing location-based social media platforms, thus allowing recommendation systems to be developed which need not rely on an extensive in-place dataset. In addition, we have shown an example of how our system could be implemented in scale to recommend food venues for different language users in different European cities (where 0.38 million food venues were analyzed using data from 26 million tweets). Finally, we have provided the results of an evaluation of the venues recommended by our approach through user evaluations.

Overall, there are several limitations to this study. First, the evaluation study was carried out only within the context of French language users and whether the results could be generalized to other language users would need to be investigated further. In addition, we focused only on Food venue recommendation and there is potential for the method proposed in this paper to be applied to other venue types (such as tourist destinations etc.). Therefore, in our future work, we would look to expand our recommendation approach to other types of venues apart from restaurants (e.g., parks and recreational, government service points etc.) and in other geographical areas (e.g., Asia and America, etc.). Another limitation is that the current approach proposed in this paper used all tweets within the vicinity of the store to represent general user experiences in that store. In the future, we are looking to develop a method which uses deep learning natural language processing techniques to better identify and match tweets that are more relevant to each corresponding venue to augment the dataset in our system.

Acknowledgements This work was partially supported by JSPS KAKENHI Grant Numbers 16H01722, 17K12686, 17H01822, 19K12240, and 19H04118.

References

1. A. Perrin, Social media usage: 2005–2015 (2015)
2. T. Hu, R. Song, Y. Wang, X. Xie, J. Luo, Mining shopping patterns for divergent urban regions by incorporating mobility data, in *Proceedings of the 25th ACM International on Conference on Information and Knowledge Management* (2016), pp. 569–578. https://doi.org/10.1145/2983323.2983803
3. T. Sakaki, M. Okazaki, Y. Matsuo, Tweet analysis for real-time event detection and earthquake reporting system development. IEEE Trans. Knowl. Data Eng. **25**(4), 919 (2013)
4. A. Sarker, K. O'Connor, R. Ginn, M. Scotch, K. Smith, D. Malone, G. Gonzalez, Social media mining for toxicovigilance: automatic monitoring of prescription medication abuse from Twitter. Drug Saf. **39**(3), 231 (2016)
5. B.J. Jansen, M. Zhang, K. Sobel, A. Chowdury, Twitter power: tweets as electronic word of mouth. J. Assoc. Inf. Sci. Technol. **60**(11), 2169 (2009)
6. J. Bao, Y. Zheng, M.F. Mokbel, Location-based and preference-aware recommendation using sparse geo-social networking data, in *Proceedings of the 20th International Conference on Advances in Geographic Information Systems*, pp. 199–208 (2012)
7. F. Liu, H.J. Lee, Use of social network information to enhance collaborative filtering performance. Expert Sys. Appl. **37**(7), 4772 (2010)
8. P. Siriaraya, Y. Nakaoka, Y. Wang, Y. Kawai, A food venue recommender system based on multilingual geo-tagged tweet analysis, in *2018 IEEE/ACM International Conference on Advances in Social Networks Analysis and Mining (ASONAM)* (2018), pp. 686–689

9. G. Dong, W. Yang, F. Zhu, W. Wang, Discovering burst patterns of burst topic in Twitter. Comput. Electr. Eng. **58**(C), 551 (2017)
10. N. Günnemann, J. Pfeffer, Finding non-redundant multi-word events on Twitter, in *Proceedings of the 2015 IEEE/ACM International Conference on Advances in Social Networks Analysis and Mining 2015* (2015), pp. 520–525
11. R. Nugroho, W. Zhao, J. Yang, C. Paris, S. Nepal, Using time-sensitive interactions to improve topic derivation in Twitter. World Wide Web **20**(1), 61 (2017)
12. A. Ritter, Mausam, O. Etzioni, S. Clark, Open domain event extraction from Twitter, in *Proceedings of the 18th ACM SIGKDD International Conference on Knowledge Discovery and Data Mining* (2012) pp. 1104–1112
13. A. Culotta, Towards detecting influenza epidemics by analyzing Twitter messages, in *Proceedings of the First Workshop on Social Media Analytics* (2010) pp. 115–122
14. T. Cheng, T. Wicks, Event detection using Twitter: a spatio-temporal approach. PLoS one **9**(6), e97807 (2014)
15. M.E. Larsen, T.W. Boonstra, P.J. Batterham, B. O'Dea, C. Paris, H. Christensen, We feel: mapping emotion on Twitter. IEEE J. Biomed. Health Inform. **19**(4), 1246 (2015)
16. M. Birkin, K. Harland, N. Malleson, P. Cross, M. Clarke, An examination of personal mobility patterns in space and time using Twitter. Int. J. Agric. Environ. Inf. Syst. **5**(3), 55 (2014)
17. R. Jurdak, K. Zhao, J. Liu, M. AbouJaoude, M. Cameron, D. Newth, Understanding human mobility from Twitter. PLoS One **10**(7), e0131469 (2015)
18. Q. Yuan, G. Cong, Z. Ma, A. Sun, N.M. Thalmann, Who, where, when and what: discover spatio-temporal topics for Twitter users, in *Proceedings of the 19th ACM SIGKDD International Conference on Knowledge Discovery and Data Mining* (2013) pp. 605–613. https://doi.org/10.1145/2487575.2487576
19. F. Luo, G. Cao, K. Mulligan, X. Li, Explore spatiotemporal and demographic characteristics of human mobility via Twitter: a case study of Chicago. Appl. Geogr. **70**, 11 (2016)
20. B. Hawelka, I. Sitko, E. Beinat, S. Sobolevsky, P. Kazakopoulos, C. Ratti, Geo-located Twitter as proxy for global mobility patterns. Cartogr. Geogr. Inf. Sci. **41**(3), 260 (2014)
21. J. Kim, M. Cha, T. Sandholm, Socroutes: safe routes based on tweet sentiments, in *Proceedings of the 23rd International Conference on World Wide Web* (2014), pp. 179–182. https://doi.org/10.1145/2567948.2577023
22. Y. Qu, J. Zhang, Trade area analysis using user generated mobile location data, in *Proceedings of the 22nd International Conference on World Wide Web* (2013), pp. 1053–1064
23. I. Eleta, J. Golbeck, Multilingual use of Twitter: social networks at the language frontier. Comput. Hum. Behav. **41**, 424 (2014)
24. F. Pla, L.F. Hurtado, Language identification of multilingual posts from Twitter: a case study. Knowl. Inf. Syst. 1–25 (2016). https://doi.org/10.1007/s10115-016-0997-x
25. K.C. Raghavi, M.K. Chinnakotla, M. Shrivastava, in *WWW '15 Companion, Proceedings of the 24th International Conference on World Wide Web* (ACM, New York, 2015), pp. 853–858. https://doi.org/10.1145/2740908.2743006
26. B. Liu, *Sentiment Analysis and Opinion Mining* (Morgan & Claypool Publishers, San Rafael, 2012)
27. B. Pang, L. Lee, Opinion mining and sentiment analysis. Found. Trends Inf. Retr. **2**(1–2), 1 (2008). https://doi.org/10.1561/1500000011
28. B. Liu, Sentiment analysis and subjectivity, in *Handbook of Natural Language Processing*, 2nd edn. (Taylor and Francis Group, Boca, 2010)
29. V. Jijkoun, M. de Rijke, W. Weerkamp, Generating focused topic-specific sentiment lexicons, in *Proceedings of the 48th Annual Meeting of the Association for Computational Linguistics* (2010), pp. 585–594. http://dl.acm.org/citation.cfm?id=1858681.1858741
30. M. Araújo, P. Gonçalves, M. Cha, F. Benevenuto, iFeel: a system that compares and combines sentiment analysis methods, in *Proceedings of the 23rd International Conference on World Wide Web* (2014), pp. 75–78. http://doi.acm.org/10.1145/2567948.2577013
31. M. Araujo, J. Reis, A. Pereira, F. Benevenuto, An evaluation of machine translation for multilingual sentence-level sentiment analysis, in *Proceedings of the 31st Annual ACM*

Symposium on Applied Computing (2016), pp. 1140–1145. https://doi.org/10.1145/2851613.2851817
32. K. Rudra, S. Rijhwani, R. Begum, K. Bali, M. Choudhury, N. Ganguly, Understanding language preference for expression of opinion and sentiment: what do hindi-English speakers do on Twitter? in *Proceedings of the 2016 Conference on Empirical Methods in Natural Language Processing* (2016), pp. 1131–1141
33. G. Linden, B. Smith, J. York, Amazon.com recommendations: item-to-item collaborative filtering. IEEE Internet Comput. **7**(1), 76 (2003)
34. B. Sarwar, G. Karypis, J. Konstan, J. Riedl, in *Proceedings of the 10th International Conference on World Wide Web* (ACM, New York, 2001), pp. 285–295
35. G. Adomavicius, A. Tuzhilin, in *Recommender Systems Handbook* (Springer, Berlin, 2011), pp. 217–253
36. D. Yang, D. Zhang, Z. Yu, Z. Wang, A sentiment-enhanced personalized location recommendation system, in *Proceedings of the 24th ACM Conference on Hypertext and Social Media* (2013), pp. 119–128
37. H. Gao, J. Tang, X. Hu, H. Liu, Content-aware point of interest recommendation on location-based social networks, in *Twenty-Ninth AAAI Conference on Artificial Intelligence* (2015)
38. S. Wang, Y. Wang, J. Tang, K. Shu, S. Ranganath, H. Liu, in *Proceedings of the 26th International Conference on World Wide Web* (International World Wide Web Conferences Steering Committee, Geneva, 2017), pp. 391–400
39. D. Quercia, L.M. Aiello, R. Schifanella, A. Davies, The digital life of walkable streets, in *Proceedings of the 24th International Conference on World Wide Web* (2015), pp. 875–884
40. M.G. Wing, A. Eklund, L.D. Kellogg, Consumer-grade global positioning system (GPS) accuracy and reliability. J. For. **103**(4), 169 (2005)
41. F.R. Bentley, N. Daskalova, B. White, Comparing the reliability of Amazon mechanical turk and survey monkey to traditional market research surveys, in *Proceedings of the 2017 CHI Conference Extended Abstracts on Human Factors in Computing Systems* (2017), pp. 1092–1099. https://doi.org/10.1145/3027063.3053335
42. M. Buhrmester, T. Kwang, S.D. Gosling, Amazon's mechanical turk: a new source of inexpensive, yet high-quality, data? Perspect. Psychol. Sci. **6**(1), 3 (2011)
43. A. Levi, O. Mokryn, C. Diot, N. Taft, Finding a needle in a haystack of reviews: cold start context-based hotel recommender system, in *Proceedings of the Sixth ACM Conference on Recommender Systems* (2012), pp. 115–122

Simplifying E-Commerce Analytics by Discovering Hidden Knowledge in Big Data Clickstreams

Konstantinos F. Xylogiannopoulos, Panagiotis Karampelas, and Reda Alhajj

1 Introduction

In recent years, retail stores seem to have been largely affected by the rapid increase of electronic ecommerce. In 2017, according to [1] more than 8000 retail stores have been closed followed by some "major bankruptcies" in 2018. These include examples like Sears or Nine West that closed a large number of stores all over the United States. These big failures are allegedly connected with the fact that the specific retailers failed to adapt to the new needs of the customers who have moved towards online retailers [1] especially in Asia. In 2016, it was reported that more than $1.915 trillion were spent on e-commerce platforms [2], while it is estimated that in 2020 the same amount will reach to $3.8 trillion [3]. This is mainly attributed to the advances in technology related to online shopping such as big data analytics that help online retailers to better target their customer needs, digital literacy of the younger generations who are very familiar with the new technologies, improved personalized experience in online stores by taking into consideration previous customer interactions automatically, as well as improvement in delivery mechanisms [1]. Online retailers invest daily much more money on studying past customers' transactions aiming to improve their customer's shopping experience while attempting to increase their revenue by providing targeted recommendations to their customers. Most of the online shopping platforms collect and analyze

K. F. Xylogiannopoulos (✉) · R. Alhajj
Department of Computer Science, University of Calgary, Calgary, AB, Canada
e-mail: kxylogia@ucalgary.ca; alhajj@cpsc.ucalgary.ca

P. Karampelas
Department of Informatics and Computers, Hellenic Air Force Academy, Dekelia Air Base, Acharnes, Greece
e-mail: panagiotis.karampelas@hafa.haf.gr

M. Kaya et al. (eds.), *Putting Social Media and Networking Data in Practice for Education, Planning, Prediction and Recommendation*, Lecture Notes in Social Networks, https://doi.org/10.1007/978-3-030-33698-1_4

customers' feedback in order to build their recommendation mechanisms and target the new customers who visit their website leaving out potentially useful input from those customers who have not left any feedback [4]. In an effort to gain insight on customers behavior, scientists have tried to exploit any trace that is left to an online retail shop from the customers. Usually these traces are recorded in the form of clicks between the different links that a customer selects during a website visit. This sequence of click actions is usually recorded automatically by the websites relating the page visits with the corresponding product views producing a stream of data that are referred to in the bibliography as clickstream data [5]. In the early days of the web, this kind of data has been used to study and understand the human behavior while visiting a website trying to confirm at the same time the appropriateness of the information structure through the sequence of the clicks [6]. Progressively, clickstream data studies have found more applications as for example in attracting new customers [5], to improve user experience while visiting a website by offering high personalization [6], or trying to explain user behavior while shopping online [7].

In e-commerce, clickstream data are usually considered as a discrete ordered sequence of website page visits by a customer. Each webpage visit can be unique or a repeated visit in case the customer visits the page again. In the context of an online retail store, each webpage is usually associated to a unique product. The sequence of webpage visits that correspond to products in our case, can easily be transformed from a clickstream analysis problem to a frequent itemset detection problem. As a consequence, it is possible to take advantage of existing efficient methods and algorithms that are used for frequent itemsets detection in order to apply them in click stream analysis and extract useful insight about customers' preferences through their clicks. Frequent itemsets detection can reveal patterns to customers' behavior that can be proved extremely valuable to marketers in order to provide either precise recommendations to a new customer or to use this knowledge to plan and conduct targeted marketing campaigns.

Since the clickstream data analysis problem can be easily transformed to a sequential pattern mining problem, a plethora of existing methodologies and algorithms that deal with the specific problem can be employed. A-priori family of algorithms is the mostly used methodology to solve the frequent sequential mining problem by reading in multiple passes the available data in an attempt to identify the candidate frequent itemsets and using a minimum support to detect the most frequent itemsets [8–11]. Similarly, another family of algorithms called pattern growth algorithms split the initial data stream to smaller databases of itemsets that can be analyzed easier, and then the most frequent itemsets are found in each database separately [12–14]. Other recent approaches to tackle the same problem apply more sophisticated techniques to mine sequential frequent patterns by taking into consideration not only the sequence of appearance of the itemset, but also other parameters such as the cost or the profit generated by each item. Thus, the specific techniques can detect the itemsets of interest faster since they take into consideration additional information rather than simply the order of the itemsets [15–17]. All the above-mentioned methodologies and algorithms work well when the interest of the study is limited to the most frequent itemsets. However, when other types of business intelligence are required, as for example, the least frequent items bought together come to the attention of the marketers, unfortunately there is no methodology that can respond to such a question.

Another important issue that needs to be addressed when attempting to respond to complicated business intelligence questions related to clickstream analysis is to find a realistic dataset that can be used to test novel methodologies in order to measure their performance. The available datasets usually vary in size, distribution and other statistical parameters. However, most of them do not reflect the actual number of visits or products that a very large e-commerce website may accommodate daily or monthly. For example, the yoochoose clickstream dataset has a little more than 33 million clicks from 9.25 million sessions (visits) for approximately 53 thousand items [4]. The specific dataset regardless of how big it may appear, does not even approach the daily number of clicks or the number of products a very large online retailer may have daily as we will see later. Another larger dataset that has been used comes from the mobile phone app Whisper [18], which contains approximately 135 million clicks from 1000 thousand users. This corresponds to 100 thousand unique website visits with an average of 1350 clicks per visit. In the same publication [18], another dataset has been used from Renren, one of the largest social networks in China. The latter dataset contains 5.9 million clicks from almost 10,000 users. This is equivalent to approximately 10,000 unique visits with an average of 59 clicks per visit. Other papers have tested their proposed methodologies on datasets from e-commerce or phone records representing few hundred thousand visits, e.g., [19, 20].

Studying the traffic of big online retailers for the holiday season as it is reported in various market surveys [21, 22], it appears that the freely available datasets do not even come close to the number of visits and products that are recorded during the specific period. More specifically, as it is reported in [21], the average daily traffic to the Top 100 e-commerce sites on October amounts to 145,811,000 while during the period 1–21 November, the average traffic is slightly higher 175,337,000. On the Thanksgiving Day, the daily traffic increases to 297,539,000, and on Black Friday the number climbs even higher to 344,407,000. The daily traffic of Cyber Monday is reported in [22]. It is very close to that of Black Friday; it is equal to 303,777,000 as it can be seen in Fig. 1.

Projecting the average daily traffic for the whole month of November, the result is equal to approximately six billion transactions for the top 100 US shopping sites. November is in general considered as the most active month in terms of online purchases since people tend to make their purchases online for the Thanksgiving, Black Friday and Cyber Monday. Another interesting aspect that appears from studying the individual indices of the most popular e-commerce websites is that Amazon got more than 50% of the online traffic of the Top 100 shopping websites in 2018, followed by Walmart and eBay with percentages a little over 10% of the traffic [21].

In this chapter, building upon the customized sequential all frequent itemset detection algorithm [23] presented in [24], a new case study has been performed

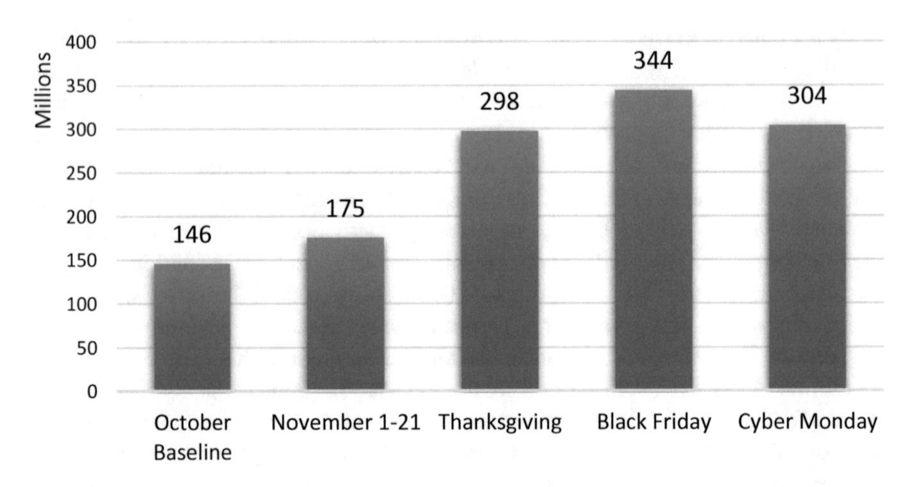

Fig. 1 Average daily traffic for the top 100 e-commerce sites in USA [21, 22]

on a composite dataset that simulates the holiday season traffic starting from November 1st ending on December 31st of the dominant e-commerce players such as Amazon, Target and Walmart. In order to simulate such a heavy traffic, the IBM Quest Synthetic Data Generator created by Agrawal and Srikant [18] has been used to create the dataset that was analyzed. The dataset simulates more than ten billion clickstreams using 650 million items. This is multiple times larger than the estimated monthly traffic of the holiday season in United States as reported in [21, 22] and presented above. The specific dataset, to the best of our knowledge, is the largest ever used in the literature for analyzing clickstream data. It outperforms by a factor of 2 the reported actual transactions in [21, 22] for the 2018 holiday season. The time needed to analyze such a big dataset in a standard desktop computer was approximately 1 week. This demonstrates how the proposed methodology can perform big data clickstream analytics in a very efficient way with limited resources.

The rest of this paper is organized as follows. Session II presents the related work in the area of clickstream analytics and sets the basis of the proposed methodology. Section 3 presents the proposed methodology, presenting the customized data structure and algorithm in order to analyze the specific dataset. Section 4 describes the details of the experiment and presents the data analysis. Finally, Sect. 5 concludes the chapter with a brief outline of the proposed methodology and the corresponding outcomes.

2 Related Work

2.1 Clickstream Analytics

Clickstream analysis has received a lot of attention from the research community since it can provide useful insight regarding customers' behavior when visiting an online store. This can help marketers to better understand their customers and work towards improving company's profit by applying targeted marketing campaigns or showing recommendations to website visitors according to their profiles. The interest of the research community to study clickstream data has led to the creation of software tools for statistical analysis packages to provide off-the-self solutions for clickstream analysis [25], though with limited applications.

The applications of clickstream analysis as previously mentioned are mainly focused on the behavior of website visitors, and as such have been studied in diverse situations. In [26], clickstream data has been used to estimate the quality of the annotation process in image labeling through crowdsource platforms. Based on the clickstream data recorded during image classification, the researcher proposed a methodology to estimate the quality of the specific classification.

In a different application, a dynamic model has been developed based on clickstream data of existing customers in a telemarketing company and it is used to predict the model of new potential customers for the company using prediction models such as logistic regression, decision tree, support vector machine, and random forest [19]. Furthermore, in [7, 18] clickstream data from real social networks are analyzed using unsupervised user behavior clustering to identify fake or dormant accounts in social networks or users who have shown hostile behavior.

In [27], the author explored clickstream data coming from the daily traffic of a shopping website in an effort to identify user behavior that can improve profitability of the shopping site. In this context, the author applied the proposed rough leader clustering methodology by identifying three interesting patterns of the customers who visited the specific website. More specifically, the dataset analyzed comprised the daily traffic in an e-commerce website in China with three million clicks that belonged to 188 thousand sessions covering 2370 product categories. While the dataset is significantly smaller than the reported holiday season traffic in the large e-commerce sites in the United States [21, 22], the author worked with even a smaller dataset composed of approximately 200 thousand clicks in 30 thousand sessions for 1823 categories. It is evident that the specific methodology cannot be applied in the magnitude of data collected from the big e-commerce players, especially during the holiday season.

In [6], the authors have analyzed clickstream data from a Danish company that wanted to know how customers perceived the use of advertising banners of their products in their pages. They also wanted to estimate the performance of the banners in terms of conversion rate. To be able to use clickstream data, the authors had to devise a new methodology to analyze them since existing models, e.g., click fact models or session fact models, could not perform well in their case. The solution was to develop a subsession model in order to improve the performance of the analysis. Nevertheless, the model developed isolates a subpart of the available information and do not work with the complete dataset due to its size.

In [28], the problem of clustering users of an online community website based on clickstream data is addressed using a graph-partitioning method. Applying the specific technique, it was possible to reduce the dimensionality of the clickstream data and thus proceed with the processing of the subsequences of data and achieve the desired concept-based clustering.

Clickstream analytics has also been used in e-learning websites where the data has been used to understand the behavior of the students and track their interaction with the different topics of the online courses [29]. The dataset was only limited to a specific online course with a certain number of student registrations, and thus it was rather feasible to analyze and come to specific conclusions.

Other researchers have used clickstream data to classify the visitors of a website into different shopping types [30]. The researchers studied approximately 30 million clickstreams composed of more than 80 million clicks coming from 11 different advertising channels such as direct traffic, search engine optimization and marketing, affiliates, social networks, email campaigns, etc. Based on their analysis, they classified the visitors of the e-commerce site into four types, namely Buying, Searching, Browsing and Bouncing that can be used by marketers to improve the conversion rate of the site.

Another contemporary research in a business retail online store used random samples of clickstream data to improve product search in an effort to increase its purchases [20]. The clickstream data used in this case are search clicks and purchase clicks. They have been used as feedback to LETORIF (LEarning TO Rank with Implicit Feedback) algorithm in order to increase the revenue of the online retail store by optimizing purchases. Finally, in another contemporary study to address the problem of complexity and high dimensionality of clickstream data analytics, a novel infrastructure architecture based on Hadoop, Spark and Kafka has been proposed, but it has not been tested [31].

As evident from the review of the related literature, clickstream analytics is a complex and resource demanding process. Existing analysis methodologies are mainly based on segmentation or clustering of data or they employ sampling to reduce the size of the data. In our case presented in this paper, we emphasize basing our analysis on the complete dataset and identifying not only frequent items visited or purchased together, but also detect products that are sold together less frequently. We believe that such an analysis can be proved very useful to marketers and provide them with insight on customer preferences and assist them in planning marketing campaigns to promote products that are sold together and avoid promotion activities of products that are not usually combined together. The specific analysis can only be performed by the customized SAFID algorithm that we already presented in [23]. The algorithm has been further expanded in this work as presented in the next section to accommodate the needs of the top 100 e-commerce stores in high season periods such as the holiday season.

2.2 Big Data Clickstream Analytics Fundamentals

Big data streams analytics usually uses clustering frameworks in order to achieve the analyses in feasible time. Systems such as Hadoop or Spark are the most frequently used with additional software tools running on top of them. In order to execute advanced methodologies and algorithms, these systems require significant infrastructures with elevated resources, which are highly costly. Moreover, the machine learning algorithms which are usually used do not guarantee high accuracy. In [24] we presented a new methodology which allows the analysis of a vast number of clickstreams with perfect accuracy. In order to prove the efficiency of the methodology we used as synthetic dataset the simulated monthly traffic of Amazon, the biggest online e-commerce retailer in the USA. However, in this paper we will present a more advanced variant of the methodology, specifically designed to analyze billions of clickstreams which represent the monthly traffic of all the e-commerce websites in the USA.

The most important feature of the proposed methodology is that it can be executed on standard desktop computers without the need of huge hardware and software infrastructure. In order to achieve this, we used data structures and algorithms specifically designed to perform exact pattern matching with perfect accuracy of the results. The data structure used is the LERP-RSA which is presented in detail in [32, 33, 35]. The LERP-RSA acronym stands for the Longest Expected Repeated Patterns Reduced Suffix Array which is an advanced variant of the classic suffix array, specifically designed for pattern detection of repeated patterns and with many more applications in other pattern detection cases [33, 35]. Based on the definition of LERP, which is the longest pattern that we expect to find in a string and occurs at least twice, and the Probabilistic Theorem of Longest Expected Repeated Pattern [33, 35], it is possible to construct a highly efficient array of all actual suffix strings by calculating the value of LERP based only on the sizes of the dataset and alphabet of the strings.

Additionally, a fundamental attribute of the LERP-RSA data structure is that it supports the classification of the dataset based on the alphabet of the strings. This way, it is possible to create many classes with very small size that can be analyzed in parallel, another important attribute of the data structure. The classification level of the data structure, as defined in [33, 35], is the exponent value of the alphabet size. Practically, this means the number of letters from the predefined alphabet of the suffix strings are used to create arrangements. For example, if the alphabet is the 26 English alphabet letters, then classification level one is $26^1 = 26$ different classes, each of them starts with a different letter of the alphabet "A" up to "Z". The importance of the classification level is that (because of the exponent definition) it can produce a vast number of classes by using small arrangements. For example, having three letters will give $26^3 = 17{,}576$ different classes starting from "AAA" and ending with "ZZZ". Therefore, for a big alphabet with a modest classification level we can create thousands of classes. This divides the total size of a dataset by a significant factor and makes it possible to run the analysis on normal hardware configuration of CPU and memory.

The algorithm used for the analysis is the All Repeated Patterns Detection (ARPaD) algorithm described in [33, 35]. So far, we have managed to use this algorithm for many different problems with outstanding results. The uniqueness of the algorithm is based on the perfect cooperation with the LERP-RSA data structure and the fact that it can detect any pattern that occurs at least twice. Therefore, ARPaD is a deterministic algorithm without the need of machine learning algorithms to be trained first. The fact that the algorithm is deterministic also guarantees that it can detect any possible repeated pattern regardless of the dataset, the pattern or the alphabet size. Furthermore, it is not only that it has the best accuracy, but it is also extremely efficient for pattern detection since its worst time complexity is $O(n \log n)$, while on average it can perform linear time complexity $O(n)$. Moreover, it has the ability to operate with the LERP-RSA classes that are stored locally or remotely in database management systems or for better performance in memory. This is of extreme importance because the efficiency of the algorithm is based on the fact that multiple and distributed classes can be analyzed in parallel. Additionally, parallelism can be executed on all classes simultaneously in one step, semi-parallel by analyzing batches of classes one after the other or in completely different time spans according to the availability of the hardware resources. Finally, another important feature of the LERP-RSA and ARPaD combination which completely diversifies them from the clustering framework-based methodologies is that the classification and parallelism process can be executed in completely different hardware and operating system units which can be completely isolated. For example, different computers running Windows or Linux can be used in combination with mobile devices using Android or iOS. This is very important because big corporations instead of using hundreds of nodes of a clustering framework, they could, for example, use the smartphones of their employees by providing computing power significantly larger and less costly.

In this chapter, a variation of the previously presented clickstream analytics methodology [24] will be presented using an advanced transformation of the LERP-RSA construction method. This new construction will allow a dataset of billions of records to be constructed on a single desktop computer without suffering any memory or disk limitation. It will show how LERP-RSA and ARPaD can be optimized to perform the best possible utilization of the available hardware in order to analyze a vast dataset in a meaningful time span.

3 The SAFID Methodology

The proposed methodology is a twofold process. It is named Sequential All Frequent Itemsets Detection (SAFID). It is based on the LERP-RSA data structure [33] which is used to store the clickstream data, and the ARPaD algorithm [32] that performs the actual analysis and identifies the frequent sequential itemset patterns as explained in the following sections.

3.1 LERP-RSA Data Structure

The LERP-RSA data structure is an advanced variation of the suffix arrays with an important diversification compared to other variants. It stores all the actual suffices of a string; not only the indices as it is the case with the traditional suffix array data structure and its alternatives. The notion behind the specific data structure is to enable the ARPaD algorithm to be executed in a highly efficient way for pattern detection on single or multiple strings. Based on the definition and construction of the LERP-RSA, it is allowed to complete the classification of the dataset based on characters of the alphabet used to construct the string(s). Moreover, as it is mentioned earlier, and it will be experimentally demonstrated later, the classification can be expanded for different levels. This offers the required scalability for big data analytics and for parallel utilization of all the available computing resources. This is very important because by using a high classification level more classes are produced with smaller sizes compared to the original dataset. Therefore, the ARPaD execution can be accelerated by a significant factor. Another advantage of the LERP-RSA algorithm is that the different classes can be stored to any storage system either locally or in distributed manner. Therefore, the process can be accelerated by limiting the read access time from a disk or a network overhead because when a class is loaded in memory and analyzed, and before the analysis of the execution is completed, the next class can be loaded in memory and, therefore, accessed directly from ARPaD without the need to wait for a disk read and/or network transfer.

3.2 ARPaD Algorithm

As mentioned earlier, the ARPaD algorithm can detect any repeated pattern that exists in a string sequence [32, 33, 35]. Furthermore, it is possible by alternating the LERP-RSA construction to achieve repeated pattern detection for a group of sequences. In such a case, ARPaD can detect not only repeated patterns between the same or different sequences, but also patterns that are repeated only because they appear once in different sequences [35]. For the specific clickstream analytics problem, we need to distinguish two different situations. The first case is when patterns appear because the user of the website moves backward and forward on the webpages during the session. Consequently, the same webpage identifiers may appear multiple times. The second case is the most obvious and difficult to be detected when the same motifs of webpage visits occur between different users. In the first case, the pattern detection process is trivial since the number of visits is usually small. However, in the second case, the pattern detection process is extremely computationally demanding and, thus, other algorithms use the notion of support to limit the results based on their frequency, which is always extremely high compared to the actual results. For example, a repeated pattern with a thousand occurrences in a dataset of one million clickstreams has support 0.00001%. Yet, an advanced, problem specific, construction of LERP-RSA allows ARPaD to be executed and detect every repeated pattern regardless of its frequency even in multivariate sequence systems.

3.3 SAFID Methodology

The classic approach of sequential itemsets detection is performed in two steps. First, the clickstream data are imported in a database system and then traditional algorithms try to detect the most frequent sequences. The second phase requires the delineation of the support, practically a frequency threshold that the algorithm will terminate when reached in order to converge to a solution in a meaningful time span. However, in our case support is meaningless since ARPaD can detect every repeated pattern regardless of its frequency. Yet, it is possible in a meta-analysis process to filter our results based on a frequency level. The important difference is that this filter can be applied as many times as required after the analysis without the need to rerun the whole process. However, there are cases where even extremely low frequency patterns are important to be detected. For example, in a periodic analysis such as daily, weekly or monthly, it is possible to detect trends on the patterns showing potential increase or reduction in the popularity of items. Moreover, low frequencies with sudden increases could show potential anomalies in the dataset mainly from spam users trying to increase or decrease the rating value of items.

Before the construction of the LERP-RSA and the execution of ARPaD, a series of transformation needs to be applied, and statistical calculation on the dataset should also be implemented. These transformations and statistical calculation are very important in order to create a more compact data structure, which will allow ARPaD to be executed in the most time and space efficient way. However, these intermediate processes do not affect the total execution time since they are executed on the fly as a new clickstream sequence enters the system. Whenever a sequence of clicks for a specific user session is recorded, it is transformed to a string. The recording of all clickstream sequences will create a multivariate system of sequences for which the LERP-RSA will be created. Finally, ARPaD will be executed and detect all repeated clickstream patterns that exist either in a single clickstream or between different clickstreams.

LERP-RSA and ARPaD require data in a specific format, and thus the original data need to be transformed to the compatible format. Thus, the SAFID methodology has been changed to incorporate five different stages. In the first stage, data preparation takes place. In the second the data are transformed to a 90-character alphabet. In the third, the data are transformed to equal length substrings and then the LERP-RSA data structure is constructed, including several possible merge sub-

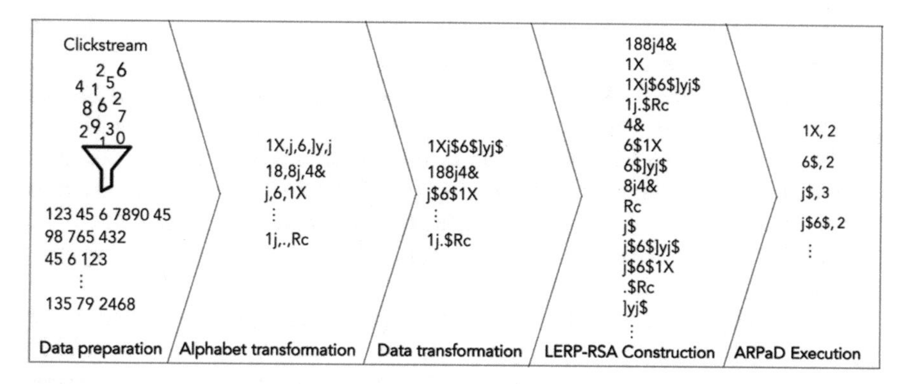

Fig. 2 The SAFID methodology flow diagram

processes. Finally, ARPaD is executed (refer to "Fig. 2"). Based on the available infrastructure, i.e., the number of available cores in the computing system, different classification levels can be defined as presented below.

The first phase is data preparation of the incoming clickstream traffic. Each clickstream is constructed from a series of indexes for each webpage visited by the user. Usually, these are derived from a decimal based numbering system. These indexes could be simple numbers, ISBN, indexes of products, merchant product or serial number, etc. An important part of the proposed methodology is the transformation of these indexes to more compact strings, using a 90 symbols alphabet which contains all decimal digits "0.9", capital "A..Z" and small "a..z" letters from the English alphabet, and symbols +−*/^!%=@#&?~_;:'.\|<>{}[](), constructing the alphabet "0123456789ABCDEFGHIJKLMNOPQRSTUVWXZYabcdefghijklmnopqrstuvw xyz+*/^!%=@#&?~_; :'.\|<>{}[]()" (double quotes not included).

Practically, the alphabet of LERP-RSA, as explained below, is the total number of itemsets and their representation, a fundamental attribute of the methodology which allows the use of alphabets of billions of letters. However, since this is not feasible with the standard symbols used in computers, the letters are constructed from basic characters and have length more than one. For example, if we have 90 items in total, then each decimal number code of the items is a letter of the alphabet used, which means that items with codes from 10 up to 90 consume two digits to represent the alphabet letter. However, if we use a 90 characters alphabet then the size is reduced to one digit per alphabet letter. This is important because in a database with millions of products indexed with a decimal system, we can reduce the length of the string codes representing the items to almost half. For example, if we have a few billion items represented with the decimal system, we need on average nine characters to represent them, while with the 90 characters alphabet, we need no more than 5.

After the transformation of the incoming itemsets codes, two very important metrics are recorded. The first one is how many clicks exist in the specific clickstream, and the second one is the classification of the web pages based on the classification level already selected. For example, if the 90 characters alphabet is used and classification level 1 has been selected then we need to construct 90 different classes for each character of the alphabet. If a product with identifier Rc is found in a clickstream then the count of class R will be incremented by 1. If another product with identifier R9 comes in a different transaction then the count of class R will be incremented by 1 again. The number of classes as mentioned earlier is very important because it helps to distribute the LERP-RSA data structure creation process to different cores in the computer's processor or in different computers and accordingly run the ARPaD algorithm in parallel to all the available computing resources.

In the second phase, the clickstream web pages data are transformed to the equivalent representation of a single string to be used as input to the ARPaD algorithm which expects item strings to have the same length. Therefore, a special character is inserted in each item identifier to match the maximum length of the longest item identifier. For example, if the number of products in a website is 8100 then the product identifiers will range from 0 to 8100, or on the 90 characters alphabet from 0 to)). Therefore, if the 45 product is clicked, first it will be converted to the new alphabet and will receive the new identifier j and then it will be transformed to j$. Here, the $ sign is the neutral character, not used in the alphabet. It is inserted in each short identifier to match the maximum length of the longest identifier. The transformed clickstream data are then combined to a single string. They are stored in the database for the ARPaD execution phase. For example, if the products 123, 45, 6, 7890, and 45 have been visited in this order, then the corresponding clickstream string will be transformed first to 1X, j, 6,]y, j, and finally to 1Xj6]yj$. The algorithm knows that length 2 is used for the items' identifiers. Therefore, it is easy to locate each item since these occur at positions $2n$ ($n = 0, 1, 2, \ldots$), and have length exactly 2.

The next phase entails the construction of the special data structure LERP-RSA that will allow the ARPaD algorithm to analyze clickstream data and detect all repeated patterns. By definition, the specific data structure receives a string and creates all suffix strings. Knowing in advance that the longest identifier as length two based on the previous transformation, it is possible to reduce the number of the suffix strings required. Instead of storing every suffix string of the clickstream string, we need to store only suffix strings that include a unique identifier of a multiplier of two. Therefore, in the previous example in which the identifiers have length 2, the suffix strings that will be created will start at position 0 with step 2. More precisely, suffix strings of the previous example will be 1Xj6]yj$, j$6$]yj$, 6$]yj$,]yj$ and j$. Any other suffix string that does not start from a $2n$ position and could be created in the traditional process of creating the LERP-RSA, such as, for example, Xj6]yj$ will be omitted and the total number of suffix strings will be 5 (as many as the products in the transactions) instead of 10.

Another important step of the LERP-RSA creation phase is sorting the suffix strings. The sorting of the LERP-RSA and of course of any array is very demanding process from memory perspective. If we use classification level 1 on a dataset with billions of clickstreams, then it is practically impossible to sort it. The advantage of LERP-RSA is to create as many smaller classes as we want with a higher classification level. However, it has to be used wisely because it could lead to heavy I/O operations and, therefore, significantly expand the execution time. For example, if classification level 4 is used on the 90 characters alphabet then 65 million classes will be created. To avoid this, we can execute the sorting on classification level 1 by optimizing it for the available memory and disk capacities. This can be done in a two stages process. First, we create a temporary class in memory up to the maximum available size of records (suffix strings), we sort it and then we save it on the disk. We continue the process by creating different files for each partially created and sorted class. When all records of the dataset are processed then we merge the already sorted partially created classes and create the full classes. This is significantly faster because the total I/O operations required is reduced.

The last phase of the analysis is executing the ARPaD algorithm on the LERP-RSA data structure prepared in the previous phase. The algorithm can run in full parallel or semi-parallel situation for the constructed classes defined in the first phase. It detects all repeated patterns, which in this case are all the repeated clicks that can be found in the clickstream data at least twice. However, depending on the available infrastructure and taking advantage of the preprocessing stages, the algorithm can be configured to run more efficiently. For example, if we have defined 90 classes for classification level 1 then 90 parallel threads of the algorithm can run simultaneously if there are 90 cores available. Usually the number of available cores is significantly smaller. Therefore, the algorithm can use the metrics calculated in the first stage, regarding the size of the classes. Smaller classes are combined together to run in the same core while classes with the largest number of clickstreams will run individually in one core. Thus, the maximum time needed for the algorithm to complete will be the time needed to analyze the class with the largest number of clickstreams.

The methodology for clickstream analytics can be described in brief with the following steps:

1. Data Preparation
   (a) Convert click identifiers to 90 characters alphabet
   (b) Count number of clicks for each transaction
   (c) Classify webpages according to the first character(s) and Classification Level
2. Data Transformation of transaction using neutral the symbol $ as suffix and create fixed length substrings
3. Create LERP-RSA using as substring step the length of the substring in (2)
   (a) Create sorted subclasses of specific size
   (b) Merge subclasses to larger subclasses until the process is completed
4. Execute ARPaD and detect of all repeated clickstreams
5. Run meta-analyses of results in (4)

4 Experimental Analysis

For the experimental analysis of our methodology, we have used a desktop computer with an Intel Xeon E5-2620 v3 processor at 2.4 GHz with 12 cores, 32 GB RAM and two mechanical disks of 500 GB and 1 TB capacity. For our case study, we have chosen a dataset that simulates a total traffic beyond anything known so far. More precisely, using the IBM Quest Synthetic Data Generator (QSDG) [34] we have created a dataset of more than ten billion clickstreams simulating a traffic larger than the total traffic of the top 100 online resellers in the USA for the most demanding month, November. Despite that, this project has been abandoned by IBM, in our case it has been proven very useful because the dataset created by the data generator is deterministic. Therefore, the dataset is reproducible, and the results are directly comparable. Moreover, it can produce a dataset that can simulate a website traffic with a good approximation by providing an average behavior of the user. Although, the majority of online users spend time searching for a specific product or something similar, it is possible that in each visit a user may go back and forth and revisit the same pages. Therefore, we care to have distribution not similar to Normal with a long tail instead of a uniform distribution of page visits (Fig. 3).

One problem we faced with IBM QSDG is that it cannot produce the size of the dataset we require. However, the fact that the QSDG operates in a deterministic way facilitates bypassing this problem by creating ten different datasets with similar parameters, except the number of items (Fig. 4). For number of items parameter, we have used ten different values that create ten completely different datasets. The value has the form 650xx, where xx starts with 00 for 65,000 up to 90 for 65,090. This

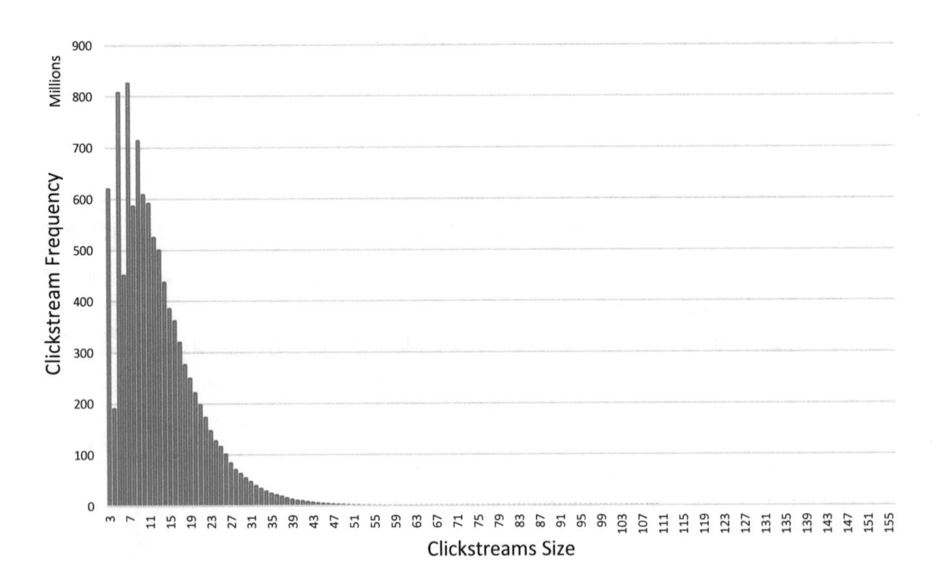

Fig. 3 Number of customer website visits per web page visits

```
seq_data_generator seq -ncust 1320000 -nitems 650xx -slen 5 -tlen 2 -ascii
```

Fig. 4 IBM quest synthetic data generator script

Table 1 IBM QSDG produced file statistics

QSDG class 650xx	Clickstreams	Clicks
00	1, 013, 764, 607	12, 856, 521, 646
10	1, 013, 775, 462	12, 856, 457, 572
20	1, 013, 745, 415	12, 856, 010, 386
30	1, 013, 758, 287	12, 855, 918, 899
40	1, 013, 757, 859	12, 855, 866, 343
50	1, 013, 757, 859	12, 855, 866, 343
60	1, 013, 754, 244	12, 855, 908, 938
70	1, 013, 767, 762	12, 855, 787, 234
80	1, 013, 728, 138	12, 855, 461, 120
90	1, 013, 739, 983	12, 855, 750, 105
Total	10, 137, 549, 616	128, 559, 548, 586

way, we do not create datasets that have exactly the same identifiers for items. They are completely different and, therefore, the same holds for the clickstreams.

The ten scripts produce in total a dataset of 10,137,549,616 clickstreams (Table 1) of length at most 156 items, from a database of 65 million items. In total, 128,559,548,586 clicks (Table 1), or item visits, have been produced from all clickstreams with approximately 13 clicks per transaction. The scripts create ASCII files that hold the clickstreams, but they are not in the proper format for our analysis. Therefore, we need to read the script and convert each clickstream to our format using the 90 characters alphabet, the neutral character, and with length 4 per item. The reason that we use length 4 per item is because with the alphabet transformation we can have up to 65,610,000 items. If we do not use the 90 characters alphabet then we need length 8 per item, which is almost double. Therefore, our alphabet transformation considerably reduces the size of the dataset.

Because of the enormous dataset size and the limited hardware resources, a direct full parallel execution is not possible. Therefore, a semi-parallel execution has been used that took advantage of the pre-statistical analysis mentioned in Sect. 3. By recording the number of occurrences per alphabet letter, we can observe in Table 2 that the occurrences per character are not uniformly distributed. For example, character “1” class is the most frequent with more than 18.5 billion occurrences while class “~” is the less frequent with approximately 350 million occurrences. This facilitates performing a semi parallel execution in ten rounds by creating different sub-classes based on the characters frequencies as shown in Table 2.

As we can observe in Table 3, the total execution time for our methodology is approximately 171 h or 7 days and 3 h. Although someone could claim that this is a large time span, first, the dataset simulates a monthly traffic. Therefore, we are four times faster than the dataset production. Second, with our limited hardware resources we had to execute the process in ten rounds. If better hardware was

Table 2 Items classes occurrences per alphabet character

Character	Occurrences	Character	Occurrences	Character	Occurrences
0	0	U	635,275,901	y	535, 783, 641
1	18, 618, 035, 401	V	768,060,609	z	553, 865, 678
2	14, 803, 504, 088	W	832,062,950	+	441, 997, 045
3	9, 344, 703, 226	X	748,970,348	−	602, 164, 183
4	5, 909, 849, 670	Z	761,655,625	*	505, 955, 686
5	3, 631, 568, 889	Y	779,358,403	/	465, 260, 173
6	2, 354, 483, 965	a	985,452,406	^	704, 125, 841
7	1, 668, 768, 379	b	939,156,993	!	536, 424, 872
8	1, 009, 485, 342	c	755,429,954	%	587, 934, 784
9	983, 248, 335	d	723,044,447	=	613, 205, 602
A	983, 090, 963	e	794,031,896	@	696, 340, 906
B	943, 620, 670	f	725,343,663	#	392, 573, 748
C	1, 005, 574, 678	g	676,592,959	&	451, 033, 255
D	953, 905, 440	h	633,236,628	?	572, 454, 940
E	937, 722, 450	i	676,853,572	~	350, 108, 132
F	674, 766, 632	j	760,703,055	_	505, 761, 224
G	1, 197, 327, 765	k	892,572,218	;	514, 248, 986
H	1, 021, 756, 825	l	733,457,963	:	523, 220, 863
I	1, 037, 345, 149	m	815,290,574	’	456, 199, 918
J	979, 979, 518	n	527,909,384	.	555, 396, 490
K	876, 432, 080	o	373,675,570	\	649, 676, 073
L	805, 781, 952	p	479,375,015	|	381, 383, 544
M	800, 440, 209	q	635,923,730	<	493, 477, 554
N	671, 225, 524	r	523,941,040	>	531, 311, 991
O	795, 027, 702	s	745,434,064	{	596, 980, 002
P	788, 313, 295	t	572,507,074	}	454, 147, 568
Q	554, 107, 634	u	634,824,755	[	494, 204, 854
R	668, 384, 452	v	558,904,829	]	440, 666, 681
S	965, 258, 741	w	586,781,963	(	441, 241, 939
T	884, 714, 327	x	569,802,686	)	16, 798, 349, 739

available that could allow us the full parallel execution then the total time could be around 18 h, which is significantly faster than the data production. Anyway, the most important outcome of this experiment is the proof of concept and technology of our methodology. Using simple, non-expensive, hardware but advanced, clever, data structure, algorithm and methods it is possible to analyze enormous datasets.

In Table 4, we can observe the vast number of patterns detected and the total occurrences per pattern length. In total, more than 27.6 billion patterns have been detected with a total number of occurrences close to 600 billion.

In Table 5, we can observe the most frequent patterns per pattern length. This is important for reproducibility purposes. Since the dataset produced by the IBM QSDG is deterministic, it is possible for anyone to recreate the exact same datasets,

Table 3 Execution time per round

Round	Alphabet start	Alphabet finish	LERP-RSA create and sort time	LERP-RSA merge time	ARPaD time	Round total time
1	1	1	08:38:52	06:27:42	05:04:25	20:10:59
2	2	2	08:55:40	06:16:07	05:36:38	20:48:25
3	3	3	07:47:24	04:27:27	04:22:49	16:37:40
4	4	5	08:40:58	03:39:07	04:04:22	16:24:27
5	6	15	09:21:43	04:59:45	02:24:01	16:45:29
6	16	27	08:10:09	03:45:36	01:40:48	13:36:33
7	28	43	08:55:10	04:16:51	01:51:41	15:03:42
8	44	64	09:16:50	04:49:14	02:08:11	16:14:15
9	65	88	08:32:06	04:22:18	02:13:42	15:08:06
10	89	89	08:46:48	06:18:49	05:10:36	20:16:13
Total time			87:05:40	49:22:56	34:37:13	171:05:49

Table 4 Pattern detected and total occurrences per pattern length

Pattern length	Number of patterns	Total occurrences
1	83, 718	128, 490, 712, 410
2	245, 669, 577	118, 395, 447, 613
3	2, 134, 822, 127	102, 488, 334, 035
4	4, 183, 836, 518	81, 792, 154, 044
5	4, 973, 472, 521	58, 800, 228, 370
6	4, 877, 199, 825	39, 879, 455, 876
7	3, 852, 260, 683	25, 668, 038, 131
8	2, 713, 689, 368	15, 885, 324, 773
9	1, 800, 226, 709	9, 698, 039, 796
10	1, 149, 879, 519	5, 755, 478, 542
11	713, 759, 415	3, 360, 048, 854
12	430, 764, 839	1, 901, 669, 447
13	251, 958, 047	1, 053, 589, 703
14	142, 695, 201	561, 531, 121
15	78, 451, 193	291, 419, 176
16	41, 559, 924	147, 217, 396
17	21, 091, 090	70, 681, 683
18	10, 385, 302	33, 600, 716
19	4, 930, 103	15, 371, 521
20	2, 202, 952	6, 664, 473
21	959, 000	2, 812, 786
22	403, 257	1, 151, 482
23	159, 224	448, 751
24	60, 549	159, 284
25	22, 344	56, 266
26	8153	19, 225
27	2767	6346
28	943	2140
29	330	718
30	96	201
31	24	50
32	6	12
33	1	2
Total	27, 630, 555, 325	594, 299, 664, 943

combine them into one large dataset and, thus, execute any current or future other pattern matching algorithm. The results of such reproducibility analysis should discover the exact same patterns. This is presented in Table 5 and was found by the LERP-RSA data structure and ARPaD algorithm which are the core mechanisms of the SAFID methodology.

Table 5 Most frequent patterns per pattern length

<table>
<tr><th>Pattern length</th><th>Total patterns</th><th>Max occurrences</th><th>Patterns</th></tr>
<tr><td>1</td><td>1</td><td>17, 395, 650, 973</td><td>1$$$</td></tr>
<tr><td>2</td><td>1</td><td>1, 104, 324, 999</td><td>3$$$1$$$</td></tr>
<tr><td>3</td><td>1</td><td>191, 562, 947</td><td>1$$$)Op91$$$</td></tr>
<tr><td>4</td><td>1</td><td>50, 617, 897</td><td>1$$$)Op91$$$)Op9</td></tr>
<tr><td>5</td><td>1</td><td>19, 690, 774</td><td>1$$$)Op91$$$)Op91$$$</td></tr>
<tr><td>6</td><td>1</td><td>5, 345, 885</td><td>1$$$)M8.1$$$)M8.1$$$)M8.</td></tr>
<tr><td>7</td><td>1</td><td>1, 774, 203</td><td>1$$$)M8.1$$$)M8.1$$$)M8.1$$$</td></tr>
<tr><td>8</td><td>1</td><td>678, 858</td><td>1$$$)P&J1$$$)P&J1$$$)P&J1$$$)P&J</td></tr>
<tr><td>9</td><td>1</td><td>280, 693</td><td>4$$$1$$$k+mw1$$$TM4R1$$$h|Xu1$$$HP]T</td></tr>
<tr><td>10</td><td>1</td><td>252, 425</td><td>1$$$)M8.1$$$\Wh?1$$$aWFK1$$$)M8.1$$$)M8.</td></tr>
<tr><td>11</td><td>1</td><td>221, 971</td><td>5$$$1$$$)M8.1$$$\Wh?1$$$aWFK1$$$)M8.1$$$)M8.</td></tr>
<tr><td rowspan="2">12</td><td rowspan="2">2</td><td rowspan="2">110, 083</td><td>2$$$O?_Q)K:=1$$$u:Xr2$$$4J#c^BB>1$$$)K:=1$$$BfvP</td></tr>
<tr><td>2$$$O?ff)Jux1$$$u;Uy2$$$4J^1^9&t1$$$)Jux1$$$Bfg4</td></tr>
<tr><td>13</td><td>1</td><td>86, 190</td><td>1$$$EK|q1$$$GLS|4$$$j;{D*k]S’06;)M8.1$$$[{{/1$$$)M8.</td></tr>
<tr><td>14</td><td>1</td><td>63, 467</td><td>5$$$2$$$7MjR/Y011$$$=zb^2$$$DtT$e@^E1$$$9[n82$$$R853n?{r</td></tr>
<tr><td>15</td><td>1</td><td>51, 268</td><td>6$$$3$$$we{gw}k7)Op91$$$)Op91$$$VnY&1$$$)Op91$$$Bu)31$$$%43R</td></tr>
<tr><td>16</td><td>1</td><td>42, 520</td><td>6$$$1$$$EK|q1$$$GLS|4$$$j;{D*k]S’06;)M8.1$$$[{{/1$$$)M8.1$$$ew?P</td></tr>
<tr><td>17</td><td>1</td><td>16, 830</td><td>6$$$1$$$t9EQ2$$$33o\i?u72$$$53@/aKxa1$$$1K7;2$$$H4U/q(b/2$$$wx&D-e7!</td></tr>
<tr><td rowspan="2">18</td><td rowspan="2">2</td><td rowspan="2">10, 242</td><td>8$$$1$$$s2L%2$$$}Oi%)Jux1$$$)Jux1$$$Obmc1$$$<cO21$$$^&EE1$$$63vn1$$$;07p</td></tr>
<tr><td>8$$$1$$$s3Hr2$$$}QZp)K:=1$$$)K:=1$$$Ob|E1$$$<e9K1$$$^?f:1$$$63/U1$$$;1*m</td></tr>
<tr><td>19</td><td>1</td><td>5, 951</td><td>2$$$O~rX)NT)1$$$u.i92$$$4J<h^Dv*1$$$)NT)1$$$Bf}o1$$$N\._2$$$pfOT)NT)1$$$)NT)</td></tr>
</table>

(continued)

Table 5 (continued)

Pattern length	Total patterns	Max occurrences	Patterns
20	1	5, 791	8$$$2$$$O~rX)NT)1$$$u.i92$$$4J<h^Dv$_*$1$$$)NT)1$$$Bf}o1$$$N\._2$$$pfOT)NT)1$$$)NT)
21	1	2, 729	8$$$1$$$:RGw1$$$1$_*$N$_*$2$$$EEFk]}bi1$$$kRd]1$$$)E+J3$$$Q~p+^0cF#2hA1$$$)E +J2$$$D2HsO{oX
22	1	933	1$$$6$_*$C84$$$4clLVcQQs4Wg)F>T1$$$lfaI1$$$)F>T2$$$L#$_*$Zcc{x1$$$)F>T1$$$jf%h1$$$HI; I1$$$]@{6
23	2	887	9$$$1$$$6$_*$bH4$$$4c$_*$7Vd^ds7Lw)Jux1$$$lh/s1$$$)Jux2$$$L&u=ce!t1$$$)Jux1$$$jh]X1$$$ HJq%1$$$];#B
			9$$$1$$$6$_*$k84$$$4c=:VeLAs8Iv)K:=1$$$liky1$$$)K:=2$$$L&}+cfXQ1$$$)K:=1$$$ji-#1$$$HJ_n1$$$]'/o
24	1	282	B$$$2$$$nCkr)F>T1$$$)F>T1$$$dkF91$$$ss}51$$$D08F1$$$)F>T1$$$_mer1$$$)F>T1$$$) F>T1$$$bbr[1$$$zT?{
25	1	130	A$$$1$$$)IYn2$$$–ya)IYn1$$$9xX>2$$$u$_*$]f!/^l1$$$PoW!1$$$#'qf1$$$V|:~1$$$) IYn2$$$ZnV+uQ[x2$$$s@N6y36?
26	2	17	A$$$1$$$)M8.3$$$Swm].Kcd)M8.1$$$O[RN1$$$X.fl1$$$T||O1$$$Ko]X3$$$3Jw{u@] Oz<9K1$$$)M8.1$$$rrd=2$$$Y\uK+ ?$_*$v
			A$$$1$$$)Op93$$$SxbW.N$_*$&)Op91$$$O]4S1$$$X\k?1$$$T< ?v1$$$Kppk3$$$3J^/u&) \z{bj1$$$)Op91$$$rtQ12$$$Y|/&+;6}
27	5	9	3$$$9/#NO&$_*$j)HEd3$$$u~JRyz\')HEd3$$$4Jt3^7MB(3zv3$$$2L)KdNfT)HEd2$$$7sY (BfDf3$$$N.Bku/6b)HEd3$$$M/{Jpa]#)HEd
			3$$$9/><O?8.)IYn3$$$u_L-y+)|)IYn3$$$4Jyo^8li(5sp3$$$2M2TdO8a)IYn2$$$7sjvBfR w3$$$N.f'u^5W)IYn3$$$M^O7pb.l)IYn
			3$$$9/wyO&Xr)F>T3$$$u?Cryy%8)F>T3$$$4Jnk^5>P(1^l3$$$2L[JdM_f)F>T2$$$7sQ [Be(h3$$$N'@Su$_*$6:)F>T3$$$M/t)pa7R)F>T
			7$$$3$$$9/><O?8.)IYn3$$$u_L-y+)\|)IYn3$$$4Jyo^8li(5sp3$$$2M2TdO8a)IYn2$$$7sjvBfR w3$$$N.f'u^5W)IYn3$$$M^O7pb.l
			7$$$3$$$9/k6ff7R_z\j3$$$h8f6yxv#)E+J2$$$])?.)E+J4$$$2L>rD{r.dMJ[uur62$$$7sH:)E +J3$$$K<Eou-44)E+J2$$$M/P)q5r_

<table>
<tr><td>28</td><td>1</td><td>9</td><td>7$$$3$$$9/><O?8.)IYn3$$$u_L-y+)|)IYn3$$$4Jyo^8li(5sp3$$$2M2TdO8a)IYn2$$$7sjvBfRw3$$$N.f'u^5W) IYn3$$$M^O7pb.l)IYn</td></tr>
<tr><td>29</td><td>1</td><td>6</td><td>8$$$3$$$O&Xrr)Ur)F>T3$$$u?Cr}I}o)F>T3$$$4Jnk^5>P)F>T2$$$Oae*)F>T2$$$Be(h<W}R2$$$N' @S^%KQ3$$$63Y3pa7R) F>T2$$$_}Z|)F>T</td></tr>
<tr><td>30</td><td>1</td><td>4</td><td>8$$$2$$$K;/R:czA2$$$1*c:FR}:3$$$EFrfKfzA(8@m3$$$kW^7v>SU)NT)2$$$QT%l)NT)4$$$Q:X:^9Z@#CeB)NT) 2$$$@vAw)NT)3$$$D3inO]GG)NT)</td></tr>
<tr><td rowspan="2">31</td><td rowspan="2">2</td><td rowspan="2">3</td><td>8$$$2$$$h-vZ:RGw3$$$1*N*bpnJsdsa4$$$DXn2EEFk]}bi)E+J2$$$kRd])E+J2$$$6NK^)E+J4$$$8N): Q~p+^0cF#2hA2$$$~I9))E+J3$$$D2HsO{oXwRFI</td></tr>
<tr><td>8$$$3$$$It([O&*j)HEd2$$$H.X_u~JR4$$$3^5j4Jt3X8=M^7MB2$$$y#B>)HEd3$$$BfDfoTSW)HEd2$$$1cy$N. Bk3$$$X@H0pa]#)HEd3$$$[6@K[kiJ)HEd</td></tr>
<tr><td rowspan="6">32</td><td rowspan="6">6</td><td rowspan="6">2</td><td>2$$$nEE3)IYn2$$$HZzC)IYn2$$$dld=)IYn2$$$su.&.g}j3$$$D0f*lDQ/l|sg2$$$YkT*)IYn2$$$_pu.<OR!3$$$srjf>lGs) IYn2$$$/Q2X)IYn2$$$bc%r)IYn</td></tr>
<tr><td>8$$$4$$$(xF$O~['^ho;)Op92$$$Qdf\u\fZ3$$$4J]<^E.#)Op93$$$7@)NHSJj)Op93$$$Bg9-L60r(|s:3$$$1JfjN |L~r2eZ3$$$P0rJpgB))Op92$$$dFqu)Op9</td></tr>
<tr><td>8$$$4$$$(yd$O_T[^i@a)P&J2$$$Qd:#u|hf3$$$4K3A^GDU)P&J3$$$7#9MHSgf)P&J3$$$BgPOL6SS(>g73$$$1Jg >N|ryr3U|3$$$P0[/ph3g)P&J2$$$dGJO)P&J</td></tr>
<tr><td>9$$$2$$$Y{Ku)NT)2$$$HYy/X:2}3$$$A_+;VS(d)NT)2$$$dq)&.nug3$$$lGW>l{^tsa6c2$$$Y8EUYmw82$$$W)a) <VW)3$$$svc>>sY2)NT)3$$$FyWW/VD1)NT)</td></tr>
<tr><td>A$$$2$$$d0l<)HEd2$$$HZcn)HEd2$$$9FQ|)HEd2$$$.fV|)HEd2$$$lCk\l\&h2$$$Yj=ac?t92$$$q5i|<MrK3$$$6) NWsqo7>jS}2$$$YA]O/O=H2$$$76(~)HEd</td></tr>
<tr><td>A$$$2$$$nEE3)IYn2$$$HZzC)IYn2$$$dld=)IYn2$$$su.&.g}j3$$$D0f*lDQ/l|sg2$$$YkT*)IYn2$$$_pu.<OR!3$$$srjf>lGs) IYn2$$$/Q2X)IYn2$$$bc%r</td></tr>
<tr><td>33</td><td>1</td><td>2</td><td>A$$$2$$$nEE3)IYn2$$$HZzC)IYn2$$$dld=)IYn2$$$su.&.g}j3$$$D0f*lDQ/l|sg2$$$YkT*)IYn2$$$_pu.<OR!3$$$srjf>lGs) IYn2$$$/Q2X)IYn2$$$bc%r)IYn</td></tr>
</table>

5 Conclusion

In this chapter, we presented an advanced analysis for big data clickstream simulating the most demanding period of e-commerce holiday season in the USA. The specific experimental case study of clickstream data analytics is based on a customized SAFID methodology variant, presented for first time, and which is able to analyze a dataset of any size with standard hardware resources. As it has been discussed, clickstream analytics is a special case of the sequential frequent itemsets detection problem. Thus, the SAFID methodology can be applied to extract useful knowledge for marketers or strategists of an online retail store. The current study simulates a multiple dataset of the monthly traffic of the top hundred U.S. e-commerce websites, which is estimated to be approximately five billion unique visits for the month of November for a 65 million products database.

Furthermore, it was proved that SAFID can be used for big data clickstream analytics regardless of the dataset size and hardware resources, since we managed to analyze more than ten billion clickstreams simulating a dataset larger than the top hundred holiday season monthly website traffic, in just 7 days with the use of a standard desktop computer while a more advanced configuration could reduce the execution time to few hours. To the best of our knowledge, there is no such case study reported in the literature so far. Moreover, we discussed how our methodology can outperform other clustered framework systems and machine learning algorithms. Therefore, the SAFID methodology indirectly proved that it can be used for even larger datasets constructed from hundreds of billions of unique website visits and products. Our target is to continue along this direction to test with larger datasets and hence further demonstrate the scalability of the proposed methodology by breaking our own records.

References

1. C. Comben, The retail apocalypse and its knock-on effects on society (2018), https://www.moneymakers.com/the-retail-apocalypse-and-its-knock-on-effects-on-society/. Accessed 2 Apr 2019
2. eMarketer, Worldwide retail ecommerce sales will reach $1.915 trillion this year (2016), https://www.emarketer.com/Article/Worldwide-Retail-Ecommerce-Sales-Will-Reach-1915-Trillion-This-Year/1014369. Accessed 20 May 2018
3. Juniper Research, Online physical goods sales to account for 13% of $30 trillion retail market by 2020 (2018), https://www.businesswire.com/news/home/20180409005544/en/Juniper-Research%2D%2D-Online-Physical-Goods-Sales/. Accessed 20 May 2018
4. T.N. Chandramohan, B. Ravindran, A neural attention based approach for clickstream mining, in *Proceedings of the ACM India Joint International Conference on Data Science and Management of Data* (ACM, 2018), pp. 118–127
5. A.L. Montgomery, S. Li, K. Srinivasan, J.C. Liechty, Modeling online browsing and path analysis using clickstream data. Mark. Sci. **23**(4), 579–595 (2004)
6. J. Andersen, A. Giversen, A.H. Jensen, R.S. Larsen, T.B. Pedersen, J. Skyt, Analyzing clickstreams using subsessions, in *Proceedings of the 3rd ACM International Workshop on Data Warehousing and OLAP* (ACM, 2000), pp. 25–32

7. G. Wang, X. Zhang, S. Tang, H. Zheng, B.Y. Zhao, Unsupervised clickstream clustering for user behavior analysis, in *Proceedings of the 2016 CHI Conference on Human Factors in Computing Systems* (ACM, 2016), pp. 225–236
8. R. Agrawal, R. Srikant, Mining sequential patterns. ed. by P.S. Yu, A.S.P. Chen, in *11th International Conference on Data Engineering (ICDE'95)* (IEEE Computer Society Press, Taipei, 1995), pp. 3–14
9. R. Srikant, R. Agrawal, *Mining Sequential Patterns: Generalizations and Performance Improvements* (Springer, Berlin, 1996), pp. 1–17
10. M.N. Garofalakis, R. Rastogi, K. Shim, SPIRIT: sequential pattern mining with regular expression constraints, in *VLDB*, vol. 99 (1999), pp. 7–10
11. M. Zhang, B. Kao, C.L. Yip, D. Cheung, A GSP-based efficient algorithm for mining frequent sequences, in *Proceedings of IC-AI* (2001), pp. 497–503
12. J. Han, J. Pei, B. Mortazavi-Asl, Q. Chen, U. Dayal, M.C. Hsu, FreeSpan: frequent pattern-projected sequential pattern mining, in *Proceedings of the Sixth ACM SIGKDD International Conference on Knowledge Discovery and Data Mining* (ACM, 2000), pp. 355–359
13. J. Pei, J. Han, B. Mortazavi-Asl, H. Pinto, Q. Chen, U. Dayal, M.C. Hsu, Prefixspan: mining sequential patterns efficiently by prefix-projected pattern growth, in *2013 IEEE 29th International Conference on Data Engineering (ICDE)* (IEEE Computer Society, 2001), pp. 0215–0215
14. M. Seno, G. Karypis, Lpminer: an algorithm for finding frequent itemsets using length-decreasing support constraint, in *Data Mining. ICDM 2001, Proceedings IEEE International Conference on 2001* (IEEE, 2001), pp. 505–512
15. D.Y. Chiu, Y.H. Wu, A.L. Chen, An efficient algorithm for mining frequent sequences by a new strategy without support counting, in *Data Engineering, 2004. Proceedings of 20th International Conference on* (IEEE, 2004), pp. 375–386
16. J. Yin, Z. Zheng, L. Cao, USpan: an efficient algorithm for mining high utility sequential patterns, in *Proceedings of the 18th ACM SIGKDD International Conference on Knowledge Discovery and Data Mining* (ACM, 2012), pp. 660–668
17. M. Zihayat, C.W. Wu, A. An, V.S. Tseng, Mining high utility sequential patterns from evolving data streams, in *Proceedings of the ASE Big Data & Social Informatics* (ACM, 2015), p. 52
18. G. Wang, X. Zhang, S. Tang, C. Wilson, H. Zheng, B.Y. Zhao, Clickstream user behavior models. ACM Trans. Web (TWEB) **11**(4), 21–37 (2017)
19. T. Sun, M. Wang, L. Liang, Predictive modeling of potential customers based on the customers clickstream data: a field study, in *Industrial Engineering and Engineering Management (IEEM), 2017 IEEE International Conference on* (IEEE, 2017), pp. 2221–2225
20. L. Wu, D. Hu, L. Hong, H. Liu, Turning clicks into purchases: revenue optimization for product search in e-commerce, in *Proceedings of the 41st International ACM SIGIR Conference on Research & Development in Information Retrieval, Ann Arbor, MI, USA, July 8–12, 2018 (SIGIR'18)*, 10 pages
21. D. Sevitt, Holiday Season 2018: Thanksgiving and Black Friday Numbers Are In! (2018). https://www.similarweb.com/blog/holiday-season-2018-thanksgiving-black-friday-numbers. Accessed 2 Apr 2019
22. D. Sevitt, Holiday Season 2018: What's the Deal with Cyber Monday? (2018). https://www.similarweb.com/blog/holiday-season-2018-cyber-monday. Accessed 2 Apr 2019
23. K.F. Xylogiannopoulos, P. Karampelas, R. Alhajj, Sequential all frequent itemsets detection: a method to detect all frequent sequential itemsets using LERP-reduced suffix array data structure and ARPaD algorithm, in *Advances in Social Networks Analysis and Mining (ASONAM), 2015 IEEE/ACM International Conference on* (IEEE, 2015), pp. 1141–1148
24. K.F. Xylogiannopoulos, P. Karampelas, R. Alhajj, Clickstream analytics: an experimental analysis of the Amazon users' simulated monthly traffic, in *Advances in Social Networks Analysis and Mining (ASONAM), 2018 IEEE/ACM International Conference on* (IEEE, 2018), pp. 841–848
25. M. Scholz, R package clickstream: analyzing clickstream data with Markov chains. J. Stat. Softw. **74**(4), 1–17 (2016)

26. E. Heim, A. Seitel, J. Andrulis, F. Isensee, C. Stock, T. Ross, L. Maier-Hein, Clickstream analysis for crowd-based object segmentation with confidence. IEEE Trans. Pattern Anal. Mach. Intell. **40**(12), 2814–2826 (2018)
27. Q. Su, L. Chen, A method for discovering clusters of e-commerce interest patterns using click-stream data. Electron. Commer. Res. Appl. **14**(1), 1–13 (2015)
28. A. Banerjee, J. Ghosh, Clickstream clustering using weighted longest common subsequences, in *Proceedings of the Web Mining Workshop at the 1st SIAM Conference on Data Mining*, vol. 143 (2001), p. 144
29. Y. Sun, C. Xin, Using coursera clickstream data to improve online education for software engineering, in *Proceedings of the ACM Turing 50th Celebration Conference-China* (ACM, 2017), pp. 16–22
30. D. Schellong, J. Kemper, M. Brettel, *Clickstream Data as a Source to Uncover Con-Sumer Shopping Types in a Large-Scale Online Setting* (2016)
31. R. Hanamanthrao, S. Thejaswini, Real-time clickstream data analytics and visualization, in *Recent Trends in Electronics, Information & Communication Technology (RTEICT), 2017 2nd IEEE International Conference on* (IEEE, 2017), pp. 2139–2144
32. K. Xylogiannopoulos, P. Karampelas, R. Alhajj, Analyzing very large time series using suffix arrays. Appl. Intell. **41**(3), 941–955 (2014)
33. K.F. Xylogiannopoulos, P. Karampelas, R. Alhajj, Repeated patterns detection in big data using classification and parallelism on LERP reduced suffix arrays. Appl. Intell. **45**(3), 567–597 (2016)
34. R. Agrawal, R. Srikant, *Quest Synthetic Data Generator* (IBM Almaden Research Center, San Jose, 2009)
35. K.F. Xylogiannopoulos, *Data Structures, Algorithms and Applications for Big Data Analytics: Single, Multiple and all Repeated Patterns Detection in Discrete Sequences*. PhD thesis, University of Calgary, 2017

Event Detection on Communities: Tracking the Change in Community Structure within Temporal Communication Networks

Riza Aktunc, Ismail Hakki Toroslu, and Pinar Karagoz

1 Introduction

Figuring out the happenings of daily life through web data analysis has become a popular research subject. The results can be applied to a wide range of domains such as emergency management or activity recommendation. Earlier efforts started under the title of *topic detection and tracking (TDT)* [4] with the analysis of news related texts to extract and to track news stories. Following TDT efforts, *event detection* problem emerged with the aim of learning about event news in a timely manner [20]. The event detection studies in the literature mostly focus on analyzing textual content from social media and web data [5, 11, 15, 16, 26], with the consideration that the social media users send timely postings about events occurring around them, which provide a rich resource for event detection. Hence, the solutions basically include text processing and mining, where the recent studies use deep learning based solutions for text mining [24, 25].

In addition to posting textual messages, events are expected to trigger certain other behaviors on individuals, such as gathering or communicating more frequently, affecting *the social context and social network* among them [17]. Therefore, in this work, we study event detection problem from network point of view [18, 19], and hypothesize that analyzing and tracking changes in community structure can reveal that an event is happening. As the source of the communication network, we use Call Detail Record (CDR) data, which is a collection of logs for each call or messaging. There are previous efforts for detecting events on CDR data in the literature [7, 12, 22]. These works focus on tracking graph attributes such as degree

R. Aktunc · I. H. Toroslu · P. Karagoz (✉)
METU, Computer Engineering Department, Ankara, Turkey
e-mail: riza.aktunc@ceng.metu.edu.tr; toroslu@ceng.metu.edu.tr; karagoz@ceng.metu.edu.tr
http://www.ceng.metu.edu.tr

M. Kaya et al. (eds.), *Putting Social Media and Networking Data in Practice for Education, Planning, Prediction and Recommendation*, Lecture Notes in Social Networks, https://doi.org/10.1007/978-3-030-33698-1_5

of nodes, and several probabilistic models to detect events on the graph extracted from CDR data. Thus, the main difference and contribution of our proposed method is that, it is based on tracking the change on community structures of the networks in consecutive time periods. There are also studies that analyze social networks in a more general setting, such as focusing on network construction from raw data and search over the graph [1]. Additionally, several other studies consider social network as a time series data and study the temporal context [10]. However, such studies aim to extract patterns on user behaviour, such as patterns of malicious use of social network, rather than event detection.

In this work, we focus on the time dimension of events, and we aim to determine the time windows in which events take place. The proposed method is based on *tracking the change in the community structures* over temporal networks. A sequence of networks, such that each network is associated with a time window along the time line, is analyzed for changes in the communities in consecutive networks. As the first step, entries of CDR data are grouped in weeks, and then, each CDR collection per week is modelled as directed weighted and unweighted graphs. The nodes of the graphs correspond to phone users in CDR data, while the edges correspond to the communication between the users. The communication can be an SMS or a voice call. For each communication type, we define different graphs.

We model the *change in the community* in three different ways: change in the number of communities, change in the central nodes of the communities, and change in the members of the communities. Within each model, there are further variations such as tracking change according to the size of the community, or considering either inclusion or exclusion of the members as the change in the community. Furthermore, we model two ensemble methods combining the basic change tracking methods. The first one combines the variations of basic change tracking on the number of communities. The second one is an ensemble of the number of communities and central nodes change tracking method. The accuracy of the methods are evaluated on a benchmark data set given in [8] in terms of precision, recall and F1-measure under varying change thresholds. The results show that community change can be used as a signal of an event happening, especially with the use of the change model involving the number of communities, and the proposed ensemble models.

An earlier version of this work proposing the use of change tracking in community structure for event detection is presented in [3]. In this paper, in addition to extensions in the descriptions of the previous models, we extend the work with improved and new models for detecting the change in the community structures. The basic method on the change in the number of communities is modified and improved by revising the grouping of communities with respect to their sizes. Additionally, we present two new change detection methods, which are hybrid solutions combining the basic change tracking models. Additional experiments on performance analysis of the new models, and scalability analysis are included, as well.

The paper is organized as follows. In Sect. 2, an overview on community detection problem is given and the community method used within our approach is summarized. In Sect. 3, the proposed methods for change tracking in community structure are described. In Sect. 4, experiments conducted and their results are presented and discussed. The paper is concluded with an overview in Sect. 5.

2 Overview on Community Detection

In order to define the *community detection* problem, firstly we need to define what *community* is. However, there is no universally accepted definition of the term *community* [9]. It is basically due to the fact that not all networks have the same community structure. Although there is no universal definition of community, there is a widespread informal definition of community concept. This definition states that a community is formed by nodes that interact with each other more frequently than with the other nodes in the network. In other words, if a group of nodes have more number of internal edges that link each other in the group than the external edges this group can be labeled as a community [9, 14, 21]. There are numerous community detection algorithms in the literature [2, 14]. We selected a modularity based community detection algorithm called as dSLM [2] in order to detect communities of temporal networks as an input to our event detection method.

Modularity is a function that gives higher output values, as the inputted graph's community structure is detected better. Thus, this function can be used to measure the quality of community detection algorithms. Apart from quality measurement, modularity is used as the basis of some community detection algorithms. These algorithms, such as dSLM, aim to detect communities in a network by trying to maximize the modularity value of the network.

Modularity is based on the idea that a randomly created graph is not expected to have community structure, so comparing the graph at hand with a randomly created graph would reveal the possible community structures. This comparison is done through comparing the actual density of edges in a subgraph and the expected edge density in the subgraph if the edges in the subgraph were created randomly. This expected edge density depends on how random the edges created. This dependency is tied to a rule that defines how to create the randomness and called as null model. A null model is a copy of an original graph and it keeps some of the original graph's structural properties but it does not reflect its community structure. There can be multiple null models for a graph such that each of them keeps different structural properties of the original graph. Using different null models for the calculation of the modularity leads to different modularity calculation methods and values. The most common null model that is used for modularity calculation is the one that preserves the degree of each vertex of the original graph. With this null model, modularity is calculated as the fraction of edges that fall in the given communities minus such fraction in the null model [9, 13]. The formula of modularity is given in Eq. (1)

$$Q = \frac{1}{2m} \sum_{ij} (A_{ij} - P_{ij}) \delta(C_i, C_j) \qquad (1)$$

In this equation, m represents the total number of edges of the graph. Sum iterates over all vertices denoted as i and j. A_{ij} is the number of edges between vertex i and vertex j in the original graph. P_{ij} is the expected number of edges between vertex i and vertex j in the null model. The δ function results as 1 if the vertex i and vertex j are in the same community ($C_i = C_j$), 0 otherwise.

dSLM [2] is a dynamic and faster version of SLM algorithm [23]. SLM is a community detection algorithm that is evolved from Louvain algorithm. Louvain algorithm is a recursive algorithm which has two main steps. Before the execution of these steps, each node is assigned to a different community of its own. Then, the following steps are executed in each recursive call:

1. The algorithm uses a simple local move heuristic in order to improve the community structure. This heuristic basically moves each node from its own community to its neighbor's community and recalculates the modularity value. If the new modularity value, which means the quality, is increased, the node would be kept in the new community, else, the node would be moved back to its previous community. This process is applied to each node for each one of its neighbors in random order and thereby heuristically the quality is aimed to be increased.
2. After that, the algorithm constructs a reduced network whose nodes are the communities that are evolved in the first step. Moreover, the weights of the edges in this reduced network are calculated as the sum of weights of the edges between the nodes which reside in the corresponding two communities. Links between nodes of the same community in the old network are represented as self-links for the node that represents that community in the new reduced network. When this reduced network is fully constructed, the first step is applied recursively on this network.

The algorithm repeats the recursion until no further improvement in modularity is obtained and thereby there are no changes in the community structure [6].

SLM algorithm changes the reduced network construction step as follows:

1. It iterates over all communities that are formed in the first step to construct subnetworks, such that, each one containing nodes from only one community.
2. Then, it applies the local moving heuristic algorithm to each subnetwork separately, after assigning each node in that subnetwork to its own singleton community in the beginning.
3. After local moving heuristic constructs new community structures for each subnetwork, the SLM algorithm creates the reduced network whose nodes are the communities detected in the subnetworks. The SLM algorithm initially defines a community for each subnetwork. Then, it assigns each node to the community that is defined for the node's subnetwork. Thus, there is a community defined for each subnetwork and detected communities in subnetworks are placed under these defined communities as nodes in the reduced network.

After these, the SLM algorithm uses the reduced network as input for the next recursive call, and all the processes starts again for the reduced network. The recursion continues until a network is constructed that cannot be reduced further [23].

In [2], the static modularity optimizer is modified so that it would detect the communities in rapidly growing large networks dynamically and efficiently. The dynamic modularity optimizer dynamically detects communities in large networks by optimizing modularity and using its own historical results.

3 Tracking the Change in Community Structure for Event Detection: Basic and Ensemble Methods

In this work, we propose that the events can be detected by tracking the change in the structure of the communities within a network in consecutive time windows. The overview of the proposed method's architecture is presented in Fig. 1. As seen in the figure, the architecture is composed of three basic modules: Preprocessing, community detection and applying change tracking methods. Given a CDR data set, once these steps are applied, output is the time windows (weeks in our case) market to include event or not. Within the method, in addition to the length of the time window, the network structure and the change detection techniques may vary.

Pre-processing In the original CDR data set, each instance is composed of three attributes: date of the call, caller id and call receiver id. There are 99,633 such call records in the CDR data. Before community detection step, the data is partitioned into time windows according to the call date. For our data set, the time window is set as week, hence we partitioned our data sets into weeks. Then, for each partition (i.e. for each week), a graph is constructed from the data. In the constructed graphs, the vertices represent the ids of the participants, and the edges represent the calls between the participants. The weight of an edge represents the number of calls made between the participants that correspond to the nodes. For instance, if the participant

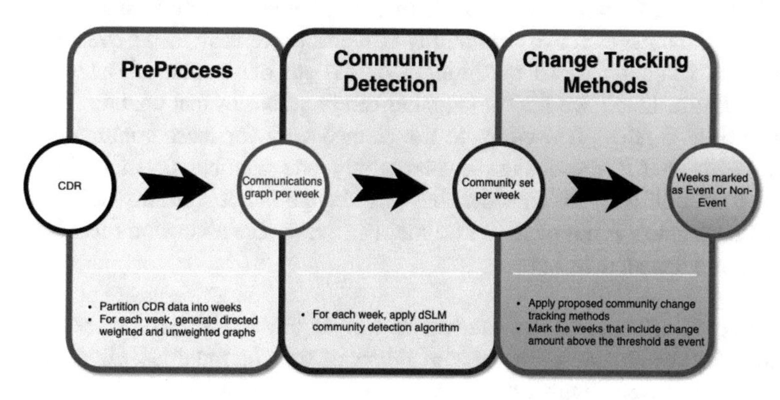

Fig. 1 The overview of the method

with id x calls the participant y z number of times in the partitioned data, then the graph of the corresponding data includes an edge between *node* x and *node* y with *weight* z.

Community Detection As the network structure, we focus on *directed weighted* and *directed unweighted* networks constructed from data partitions. As the community detection technique for both kinds of networks, we use dSLM algorithm given in [2].

Change Tracking In order to track the change, we compare the community structure parameter values against the previous time window. We propose to track the change in the community structure through three basic methods, and further develop two ensemble methods based on these basic methods. These three basic methods track the change in the parameters of *Number of Communities*, *Central Nodes*, and *Community Members*, respectively. As the first ensemble method, we combine the variations of the basic model on the change tracking in the number of communities. As the second ensemble method, we use the basic method on change in the central nodes and the change in the number of communities together. In the rest of this section, the details of these basic and ensemble methods are presented.

3.1 *Basic Methods*

Tracking the Change in the Number of Communities Our first change tracking model is defined on the parameter of the total number of communities. The change tracking mechanism is defined as follows: if the amount of change in the number of communities between two consecutive time windows exceeds a given threshold, we mark the latter window to include an event. Here, the size of the communities is an important factor to be considered. Considering all kinds of communities may lead to incorrect predictions since events may only affect communities of certain size. Hence, we focus on the changes in communities of certain size. In order to determine the most effective community size group, we analyze the event detection performance under various partitioning of the set of communities (as given in Sect. 4). For instance, we find the number of communities that contain 3–5 people in each time window, and compute the change only for these communities. The exception is that if the change involves only one member (even if the change threshold is fulfilled), we consider that this change is not a strong indication, and hence that window is not marked to contain an event. The algorithm for this method is given in Algorithm 1.

Tracking the Change in the Central Nodes In this method, for event detection, we consider the change in the *central nodes* of the communities. Given a ranked list of nodes (users) with respect to centrality score, we determine the central nodes of a community as the top 20% of the nodes in the list. We compute

Algorithm 1 Event detection via change on # of comm

```
Require: setOfComm, minCommSize, maxCommSize, changeThreshold
Ensure: events
  prev = NumberOfComm(setOfComm(t1), minCommSize, maxCommSize)
  for i=2 to timeWindowCount do
    cur = NumberOfComm(setOfComm( setOfComm(ti), minCommSize, maxCommSize)
    change = abs(cur − prev)/prev
    if change ≥ changeThreshold and abs(cur - prev) > 1 then
      add i to events
    end if
    prev = cur
  end for
  return events
```

Algorithm 2 Event detection via change on central nodes

```
Require: setOfComm, changeComptMethod, changeThreshold
Ensure: events
  prev = setOfComm(t1).centralNodes
  for i=2 to timeWindowCount do
    cur =setOfComm(ti).centralNodes
    change = ComputeChangeOnCentralNodes (
    changeCompMethod, prev, cur)
    changeInSize = abs(cur.size - prev.size)
    if change ≥ changeThreshold and changeInSize > 1 then
      add i to events
    end if
    prev = cur
  end for
  return events
```

the change of central nodes in four different ways. The simplest one is that we compute the change in the number of central nodes in consecutive time windows (denoted as SIZE). The second one is considering the change as the number of central nodes of previous window that are not central nodes any more in the current window (denoted as NOT_ANY_MORE). The third way is that we compute change as the number of central nodes of current window that were not central nodes in the previous window (denoted as NEW). The last and the most complex change computation method involves both NOT_ANY_MORE and NEW nodes, hence the addition of both of the values is used as the magnitude of the change (denoted as NOT_ANY_MORE_NEW). The algorithm for this method is presented in Algorithms 2 and 3.

Tracking the Change in the Community Members We consider the change in the community members as another candidate for event indicator. However, community detection techniques do not assign global identifiers to communities to be tracked over time windows. To be able to track the change within the same community, we assume that two communities in consecutive windows are the same

Algorithm 3 Compute change on central nodes

```
Require: changeCompMethod, prev, cur
Ensure: change
  if changeCompMethod is SIZE then
    return abs(cur. size - prev.size) / prev.size
  else if changeCompMethod is NOT_ANY_MORE then
    return notAnyMoreCentralNodeCount / prev.size
  else if changeCompMethod is NEW then
    return newCentralNodeCount / cur.size
  else if changeCompMethod is NOT_ANY_MORE_NEW then
    return (notAnyMoreCentralNodeCount +
    newCentralNodeCount) / cur.size
  end if
```

Algorithm 4 Event detection via change on comm members

```
Require: setOfComm, changeCompMethod, changeThreshold
Ensure: events
  prev = setOfComm(t1).members
  for i=2 to timeWindowCount do
    cur = setOfComm(ti).members
    change = ComputeChangeOnMembers(
    changeCompMethod, prev, cur)
    if change ≥ changeThreshold then
      add i to events
    end if
    prev = cur
  end for
  return events
```

Algorithm 5 Compute change on community members

```
Require: changeCompMethod, prev, cur
Ensure: change
  if changeCompMethod is MIN then
    return min of the change values on the comm members
  else if changeCompMethod is AVERAGE then
    return avg of the change values on the comm members
  else if changeCompMethod is MAX then
    return max of the change values on the comm members
  end if
```

if they have common central nodes. Therefore, we firstly find the communities of consecutive windows around similar central nodes. Then, we compute the change in the number of members within the community. We compute the change in three different ways: the *minimum (MIN)* of change magnitude, the *average (AVG)* of the change magnitude, and the *maximum (MAX)* change magnitude. The algorithm of the method is given in Algorithms 4 and 5.

3.2 Ensemble Methods

Combining the Variations of the Change Tracking in the Number of Communities In the first ensemble method, we hypothesize that events can be detected better if the results of the variations of the number of communities based event detection method are merged. In the basic method (given in Sect. 3.1), we hypothesize that events might arise from communities of *certain size*. Our motivation behind this hybrid approach is to extend the initial hypothesis to cover the cases that an event might arise from set of communities of certain *several* sizes. For instance an event may involve communities that contain 3–5 and 11–20 people. The initial basic method would not be able to detect such kind of events, whereas this ensemble method can detect the kind of events that involve communities of different sizes.

The ensemble method takes the communication network, range of minimum and maximum community sizes, change threshold and detection threshold parameters as input. It executes the basic event detection method based on the change tracking in the number of communities with the given parameters. If the count of events of a week fulfills the predefined detection threshold, the ensemble method marks that week to include an event. The details of the first ensemble method can be seen in Algorithm 6.

Algorithm 6 Event detection via ensemble of change on # of comm

```
Require: setOfComm, listOf(minCommSize, maxCommSize),
  changeThreshold, detectionThreshold
Ensure: events
  events = emptyList()
  listOfEventList = emptyList()
  for each minCommSize, maxCommSize in listOf(minCommSize, maxCommSize) do
    eventList = EventDetectionviaChangeon#ofComm(
    setOfComm, minCommSize, maxCommSize, changeThreshold)
    add eventList to listOfEventList
  end for
  for i=2 to timeWindowCount do
    detectionCount = 0
    for each eventList in listOfEventList do
      if eventList contains i then
        detectionCount = detectionCount + 1
      end if
    end for
    if detectionCount ≥ detectionThreshold then
      add i to events
    end if
  end for
  return events
```

Combining the Change Tracking in the Number of Communities and the Change Tracking in the Central Nodes This hybrid approach is based on the idea that an event might be related with both central node changes and number of communities changes. Therefore, we propose an ensemble method that tracks the changes both in the central nodes and in the number of communities in order to mark a week for event inclusion.

This ensemble method takes the network, the range of minimum and maximum community size, central node change detection type, change threshold and detection threshold parameters as input. The basic event detection method on the change tracking in the number of communities is applied on the data under the given input parameters. Similarly, the basic method on the change tracking in the central nodes is also executed on the same data. A week is marked to include an event, if its event detection count fulfills the given detection threshold parameter. In fact, for this ensemble method, practically the only applicable event detection thresholds are 1 and 2 in terms of event count, as each of the two methods can return either 0 or 1 as the output. In other words, we can state that count threshold value one refers to or/union, and count threshold value two refers to and/intersection of the basic methods' results. The details of this algorithm are given in Algorithm 7.

Algorithm 7 Event detection via ensemble of change on # of comm and change on central nodes

```
Require: setOfComm, minCommSize, maxCommSize, changeComptMethod
  changeThreshold, detectionThreshold
Ensure: events
  events = emptyList()
  numberOfCommEvents = EventDetectionviaChangeon#ofComm(
  setOfComm, minCommSize, maxCommSize, changeThreshold)
  centralNodeEvents = EventDetectionviaChangeonCentralNodes(
  setOfComm, changeComptMethod, changeThreshold)
  for i=2 to timeWindowCount do
    detectionCount = 0
    if numberOfCommEvents contains i then
      detectionCount = detectionCount + 1
    end if
    if centralNodeEvents contains i then
      detectionCount = detectionCount + 1
    end if
    if detectionCount ≥ detectionThreshold then
      add i to events
    end if
  end for
  return events
```

4 Experiments on Event Detection Performance

4.1 Data Set

For the evaluation of the methods, we used the *Reality Mining* data set [8]. The data set involves call detail records (CDR) of 97 faculty, student, and staff at MIT over 50 weeks. As the ground truth, the following events are marked per week: semester breaks, exam and sponsor weeks, and holidays. The data set contains logs for voice call, SMS and bluetooth activities. In our experiments, we used voice call logs, and built a sequence of temporal networks corresponding to 1-week time windows. For community detection, we used a modularity based community detection algorithm [2] (as given in Sect. 2), to find communities within each week's network.

4.2 Experiments and Results

Event detection experiments are conducted on weighted and unweighted versions of the network under varying change thresholds. The experiments of the method that detects events via tracking the change on community members have the same results for both weighted and unweighted versions of the network. Therefore, we present the results for this method as a single table.

In the tables, the first column includes the change model parameter analyzed. For basic model, this column includes the community size, the type of change in central node, or the type of change in the community members, depending on the model to be analyzed. For ensemble methods, the first column includes a set of change parameters that belong to the models that are combined. For the ensemble of number of communities methods' variations, this includes a set of community size ranges. For the ensemble of number of communities and central nodes methods, this column includes the community size range and the change detection method. In the ensemble methods, additionally, the first column includes event detection threshold value specified after *DT* prefix. In all the tables, the second column shows the applied change threshold. For each method, we present the best three results with respect to F1-measurement, per parameter setting.

Experiments on Basic Methods For the first basic method of change detection in the number of communities, initially, we grouped the communities with respect to the size by observing the number of communities of each size range on several week samples. The results for this type of partitioning are given in Tables 1 and 2 for weighted and unweighted networks, respectively. We further analyzed the effect of partitioning under several variations, and observed that the best result is obtained under the partitioning with respect to the following size ranges: 3–5, 6–8, 9–9, 10–11, 12–16, 17–31, 32–90. The results under this partitioning are given in Tables 3 and 4 for weighted and unweighted graphs, respectively. We have observed that this

Table 1 Change in the number of communities on weighted network (initial version of partitioning)

Type	Thr	Precision	Recall	F1-meas.
3–5 nodes	0.05	0.62	0.81	0.70
3–5 nodes	0.15	0.62	0.81	0.70
3–5 nodes	0.25	0.6	0.75	0.67
6–10 nodes	0.05	0.38	0.56	0.46
6–10 nodes	0.15	0.38	0.56	0.46
6–10 nodes	0.25	0.36	0.5	0.42
11–20 nodes	0.05	0.48	0.88	0.63
11–20 nodes	0.15	0.48	0.88	0.63
11–20 nodes	0.25	0.5	0.75	0.60
21–30 nodes	0.05	0.32	0.44	0.38
21–30 nodes	0.15	0.32	0.44	0.38
21–30 nodes	0.25	0.32	0.44	0.38
21–30 nodes	0.35	0.17	0.19	0.18
31–40 nodes	0.05	0.55	0.38	0.45
31–40 nodes	0.15	0.55	0.38	0.45
31–40 nodes	0.25	0.55	0.38	0.45

Table 2 Change in the number of communities on unweighted network (initial version of partitioning)

Type	Thr	Precision	Recall	F1-meas.
3–5 nodes	0.05	0.62	0.81	0.70
3–5 nodes	0.15	0.62	0.81	0.70
3–5 nodes	0.25	0.62	0.81	0.70
6–10 nodes	0.05	0.43	0.62	0.51
6–10 nodes	0.15	0.43	0.62	0.51
6–10 nodes	0.25	0.43	0.62	0.51
11–20 nodes	0.05	0.54	0.94	0.69
11–20 nodes	0.15	0.54	0.94	0.69
11–20 nodes	0.25	0.58	0.88	0.7
21–30 nodes	0.05	0.48	0.69	0.57
21–30 nodes	0.15	0.48	0.69	0.57
21–30 nodes	0.25	0.48	0.69	0.57
31–40 nodes	0.05	0.53	0.56	0.55
31–40 nodes	0.15	0.53	0.56	0.55
31–40 nodes	0.25	0.5	0.5	0.5

change in the partitioning did not lead to an increase in the maximum F1-measure, however, the average performance of the size ranges has increased. Additionally, these new size ranges provide better results in ensemble method.

As shown in Tables 5 and 6, the method of tracking change in central nodes present a similar performance for weighted and unweighted graphs. On the overall, although the recall is high, the precision value is lower than the method on the number of communities, due to high number of false positives.

Table 3 Change in the number of communities on weighted network (improved partitioning)

Type	Thr	Precision	Recall	F1-meas.
3–5 nodes	0.05	0.62	0.81	0.70
3–5 nodes	0.10	0.62	0.81	0.70
3–5 nodes	0.15	0.62	0.81	0.70
6–8 nodes	0.45	0.71	0.62	0.67
6–8 nodes	0.50	0.71	0.62	0.67
6–8 nodes	0.55	0.71	0.62	0.67
12–16 nodes	0.35	0.50	0.62	0.56
12–16 nodes	0.40	0.50	0.62	0.56
12–16 nodes	0.60	0.56	0.56	0.56
32–90 nodes	0.05	0.64	0.44	0.52
32–90 nodes	0.10	0.64	0.44	0.52
32–90 nodes	0.15	0.64	0.44	0.52
10–11 nodes	0.05	0.38	0.50	0.43
10–11 nodes	0.10	0.38	0.50	0.43
10–11 nodes	0.15	0.38	0.50	0.43
17–31 nodes	0.05	0.38	0.50	0.43
17–31 nodes	0.10	0.38	0.50	0.43
17–31 nodes	0.15	0.35	0.44	0.39
9–9 nodes	0.05	0.46	0.38	0.41
9–9 nodes	0.10	0.46	0.38	0.41
9–9 nodes	0.15	0.46	0.38	0.41

Table 4 Change in the number of communities on unweighted network (improved partitioning)

Type	Thr	Precision	Recall	F1-meas.
3–5 nodes	0.05	0.62	0.81	0.70
3–5 nodes	0.10	0.62	0.81	0.70
3–5 nodes	0.15	0.62	0.81	0.70
32–90 nodes	0.05	0.60	0.75	0.67
32–90 nodes	0.10	0.60	0.75	0.67
32–90 nodes	0.15	0.58	0.69	0.63
12–16 nodes	0.60	0.57	0.75	0.65
12–16 nodes	0.55	0.55	0.75	0.63
12–16 nodes	0.05	0.43	1.00	0.60
9–9 nodes	0.05	0.89	0.50	0.64
9–9 nodes	0.10	0.89	0.50	0.64
9–9 nodes	0.15	0.89	0.50	0.64
17–31 nodes	0.25	0.50	0.69	0.58
17–31 nodes	0.20	0.48	0.69	0.56
17–31 nodes	0.05	0.46	0.69	0.55
6–8 nodes	0.05	0.41	0.44	0.42
6–8 nodes	0.10	0.41	0.44	0.42
6–8 nodes	0.15	0.41	0.44	0.42
10–11 nodes	0.55	0.33	0.38	0.35
10–11 nodes	0.60	0.33	0.38	0.35
10–11 nodes	0.65	0.33	0.38	0.35

Table 5 Change in the central nodes within communities on weighted network

Type	Thr	Precision	Recall	F1-meas.
NEW	0.05	0.33	0.94	0.49
NEW	0.15	0.33	0.94	0.49
NEW	0.25	0.33	0.94	0.49
NOT_ANY_MORE	0.05	0.33	0.94	0.49
NOT_ANY_MORE	0.15	0.33	0.94	0.49
NOT_ANY_MORE	0.65	0.39	0.75	0.52
NOT_ANY_MORE_NEW	0.05	0.33	0.94	0.49
NOT_ANY_MORE_NEW	0.15	0.33	0.94	0.49
NOT_ANY_MORE_NEW	0.25	0.33	0.94	0.49
SIZE	0.05	0.32	0.75	0.45
SIZE	0.15	0.36	0.5	0.42
SIZE	0.25	0.27	0.19	0.23

Table 6 Change in the central nodes within communities on unweighted network

Type	Thr	Precision	Recall	F1-meas.
NEW	0.35	0.34	0.94	0.5
NEW	0.45	0.34	0.94	0.5
NEW	0.55	0.34	0.94	0.5
NOT_ANY_MORE	0.05	0.33	0.94	0.49
NOT_ANY_MORE	0.15	0.33	0.94	0.49
NOT_ANY_MORE	0.65	0.39	0.81	0.53
NOT_ANY_MORE_NEW	0.05	0.33	0.94	0.49
NOT_ANY_MORE_NEW	0.15	0.33	0.94	0.49
NOT_ANY_MORE_NEW	0.25	0.33	0.94	0.49
SIZE	0.05	0.31	0.75	0.44
SIZE	0.15	0.36	0.5	0.42
SIZE	0.25	0.25	0.19	0.22

Among the three basic change tracking methods, the one on the change in the community members has the lowest event detection performance (Table 7). Among the variations of this basic method, considering the minimum change can not capture any event, which is an expected result since small change in the community may be observed under ordinary behaviour as well. The performance under the maximum change and average change is just the same, showing a strong bias towards marking week to include an event.

Experiments on Ensemble Methods Among the basic community structure change models, since the first two perform better, we constructed two ensemble methods by combining variations of them. The first one combines the variations of the first basic method, whereas the second ensemble method combines the best performing variations of the first and the second basic methods.

Table 7 Change in the community members

Type	Thr	Precision	Recall	F1-meas.
AVG	0.05	0.33	1	0.5
AVG	0.15	0.33	1	0.5
AVG	0.25	0.33	1	0.5
MAX	0.05	0.33	1	0.5
MAX	0.15	0.33	1	0.5
MAX	0.25	0.33	1	0.5
MIN	0.05	0	0	0
MIN	0.15	0	0	0
MIN	0.25	0	0	0

Table 8 Ensemble of number of communities on weighted network

Type	Thr	Precision	Recall	F1-meas.
3–5, 6–8, 9–9, 10–11, 12–16, 17–31, 32–90, DT: 3	0.05	0.62	0.94	0.75
3–5, 6–8, 9–9, 10–11, 12–16, 17–31, 32–90, DT: 3	0.10	0.62	0.94	0.75
3–5, 6–8, 9–9, 10–11, 12–16, 17–31, 32–90, DT: 3	0.15	0.62	0.94	0.75
3–5, 6–8, 9–9, 10–11, 12–16, 17–31, 32–90, DT: 2	0.45	0.48	0.94	0.64
3–5, 6–8, 9–9, 10–11, 12–16, 17–31, 32–90, DT: 2	0.50	0.48	0.94	0.64
3–5, 6–8, 9–9, 10–11, 12–16, 17–31, 32–90, DT: 2	0.55	0.50	0.81	0.62
3–5, 6–8, 9–9, 10–11, 12–16, 17–31, 32–90, DT: 4	0.05	0.64	0.56	0.60
3–5, 6–8, 9–9, 10–11, 12–16, 17–31, 32–90, DT: 4	0.10	0.64	0.56	0.60
3–5, 6–8, 9–9, 10–11, 12–16, 17–31, 32–90, DT: 4	0.15	0.62	0.50	0.55
3–5, 6–8, 9–9, 10–11, 12–16, 17–31, 32–90, DT: 1	0.70	0.42	0.94	0.58
3–5, 6–8, 9–9, 10–11, 12–16, 17–31, 32–90, DT: 1	0.75	0.42	0.94	0.58
3–5, 6–8, 9–9, 10–11, 12–16, 17–31, 32–90, DT: 1	0.55	0.40	1.00	0.57
3–5, 6–8, 9–9, 10–11, 12–16, 17–31, 32–90, DT: 5	0.25	0.67	0.25	0.36
3–5, 6–8, 9–9, 10–11, 12–16, 17–31, 32–90, DT: 5	0.30	0.67	0.25	0.36
3–5, 6–8, 9–9, 10–11, 12–16, 17–31, 32–90, DT: 5	0.35	0.67	0.25	0.36
3–5, 6–8, 9–9, 10–11, 12–16, 17–31, 32–90, DT: 6	0.05	1.00	0.19	0.32
3–5, 6–8, 9–9, 10–11, 12–16, 17–31, 32–90, DT: 6	0.10	1.00	0.19	0.32
3–5, 6–8, 9–9, 10–11, 12–16, 17–31, 32–90, DT: 6	0.15	1.00	0.19	0.32

The results of the first ensemble method are given in Tables 8 and 9 on weighted and unweighted graphs, respectively. Similarly, the performance for weighted and unweighted versions of the graph for the second ensemble method are given in Tables 10 and 11, respectively. The event detection performance in terms of F1-measure is higher for unweighted graphs for both of the ensemble methods.

Experiments on Scalability In order to analyze the scalability of the proposed method, we conducted an experiment to measure the time performance under varying data set size. The original data set contains nearly 100,000 call detail records. For each week of this data, we constructed five different versions by constructing multiples of the data set by 1, 2, 4, 8, and 16 Each of the resulting five

Table 9 Ensemble of number of communities on unweighted network

Type	Thr	Precision	Recall	F1-meas.
3–5, 6–8, 9–9, 10–11, 12–16, 17–31, 32–90, DT: 4	0.05	0.68	0.94	0.79
3–5, 6–8, 9–9, 10–11, 12–16, 17–31, 32–90, DT: 4	0.10	0.68	0.94	0.79
3–5, 6–8, 9–9, 10–11, 12–16, 17–31, 32–90, DT: 4	0.25	0.70	0.88	0.78
3–5, 6–8, 9–9, 10–11, 12–16, 17–31, 32–90, DT: 3	0.35	0.52	0.88	0.65
3–5, 6–8, 9–9, 10–11, 12–16, 17–31, 32–90, DT: 3	0.05	0.48	0.94	0.64
3–5, 6–8, 9–9, 10–11, 12–16, 17–31, 32–90, DT: 3	0.10	0.48	0.94	0.64
3–5, 6–8, 9–9, 10–11, 12–16, 17–31, 32–90, DT: 5	0.20	0.89	0.50	0.64
3–5, 6–8, 9–9, 10–11, 12–16, 17–31, 32–90, DT: 5	0.25	0.89	0.50	0.64
3–5, 6–8, 9–9, 10–11, 12–16, 17–31, 32–90, DT: 5	0.30	0.89	0.50	0.64
3–5, 6–8, 9–9, 10–11, 12–16, 17–31, 32–90, DT: 1	0.85	0.46	1.00	0.63
3–5, 6–8, 9–9, 10–11, 12–16, 17–31, 32–90, DT: 1	0.90	0.46	1.00	0.63
3–5, 6–8, 9–9, 10–11, 12–16, 17–31, 32–90, DT: 1	0.95	0.46	1.00	0.63
3–5, 6–8, 9–9, 10–11, 12–16, 17–31, 32–90, DT: 2	0.55	0.44	1.00	0.62
3–5, 6–8, 9–9, 10–11, 12–16, 17–31, 32–90, DT: 2	0.45	0.42	1.00	0.59
3–5, 6–8, 9–9, 10–11, 12–16, 17–31, 32–90, DT: 2	0.50	0.42	1.00	0.59
3–5, 6–8, 9–9, 10–11, 12–16, 17–31, 32–90, DT: 6	0.05	1.00	0.19	0.32
3–5, 6–8, 9–9, 10–11, 12–16, 17–31, 32–90, DT: 6	0.10	1.00	0.19	0.32
3–5, 6–8, 9–9, 10–11, 12–16, 17–31, 32–90, DT: 6	0.15	1.00	0.19	0.32

different versions of data set contains nearly 100,000, 200,000, 400,000, 800,000, and 1.6M call detail records, respectively. As in the previous experiments, we performed the analysis for both weighted and unweighted graphs.

The execution times of the proposed method on the scaled data sets are presented in Table 12, Figs. 2 and 3. Table 12 lists all the durations in seconds, whereas Figs. 2 and 3 show the nature of the trend for weighted and unweighted graphs, respectively. As seen in the results, the quadratic complexity of the community detection method is reflected in the execution times.

Overview We can summarize the results of the study as follows:

- Among basic change tracking models, event detection according to the change in number of communities has the highest scores in terms of precision and F1-measure. Although it is the simplest approach, it provides a stronger indicator for event. Additionally, it has better potential to be applied online due to its lower computationally cost.
- Change tracking on smallest communities (communities with 3–5 members) provides the highest precision and recall results. This indicates that an event triggers communication among small groups that emerges new, possibly short-lived communities.
- As a general observation for the first two basic change tracking methods, lower change threshold values provides the highest accuracy scores. Except for just a

Table 10 Ensemble of number of communities and central nodes on weighted network

Type	Chng	Precision	Recall	F1-meas.
3–5, SIZE, DT: 2	0.05	0.60	0.75	0.67
3–5, NEW, DT: 2	0.05	0.60	0.75	0.67
3–5, NOT_ANY_MORE, DT: 2	0.05	0.60	0.75	0.67
3–5, NOT_ANY_MORE_NEW, DT: 2	0.05	0.60	0.75	0.67
6–8, SIZE, DT: 2	0.05	0.60	0.56	0.58
6–8, NEW, DT: 2	0.05	0.60	0.56	0.58
6–8, NOT_ANY_MORE, DT: 2	0.05	0.60	0.56	0.58
6–8, NOT_ANY_MORE_NEW, DT: 2	0.05	0.60	0.56	0.58
12–16, SIZE, DT: 2	0.05	0.45	0.62	0.53
12–16, NEW, DT: 2	0.05	0.45	0.62	0.53
12–16, NOT_ANY_MORE, DT: 2	0.05	0.45	0.62	0.53
12–16, NOT_ANY_MORE_NEW, DT: 2	0.05	0.45	0.62	0.53
32–90, SIZE, DT: 2	0.05	0.64	0.44	0.52
32–90, NEW, DT: 2	0.05	0.64	0.44	0.52
32–90, NOT_ANY_MORE, DT: 2	0.05	0.64	0.44	0.52
32–90, NOT_ANY_MORE_NEW, DT: 2	0.05	0.64	0.44	0.52
3–5, SIZE, DT: 1	0.05	0.34	1.00	0.51
3–5, NEW, DT: 1	0.05	0.34	1.00	0.51
3–5, NOT_ANY_MORE, DT: 1	0.05	0.34	1.00	0.51
3–5, NOT_ANY_MORE_NEW, DT: 1	0.05	0.34	1.00	0.51
6–8, SIZE, DT: 1	0.05	0.34	1.00	0.51
6–8, NEW, DT: 1	0.05	0.34	1.00	0.51
6–8, NOT_ANY_MORE, DT: 1	0.05	0.34	1.00	0.51
6–8, NOT_ANY_MORE_NEW, DT: 1	0.05	0.34	1.00	0.51
17–31, SIZE, DT: 1	0.05	0.34	1.00	0.51
17–31, NEW, DT: 1	0.05	0.34	1.00	0.51
17–31, NOT_ANY_MORE, DT: 1	0.05	0.34	1.00	0.51
17–31, NOT_ANY_MORE_NEW, DT: 1	0.05	0.34	1.00	0.51
12–16, SIZE, DT: 1	0.05	0.33	1.00	0.50
12–16, NEW, DT: 1	0.05	0.33	1.00	0.50
12–16, NOT_ANY_MORE, DT: 1	0.05	0.33	1.00	0.50
12–16, NOT_ANY_MORE_NEW, DT: 1	0.05	0.33	1.00	0.50
9–9, SIZE, DT: 1	0.05	0.32	0.94	0.48
9–9, NEW, DT: 1	0.05	0.32	0.94	0.48
9–9, NOT_ANY_MORE, DT: 1	0.05	0.32	0.94	0.48
9–9, NOT_ANY_MORE_NEW, DT: 1	0.05	0.32	0.94	0.48
10–11, SIZE, DT: 1	0.05	0.33	0.94	0.48
10–11, NEW, DT: 1	0.05	0.33	0.94	0.48
10–11, NOT_ANY_MORE, DT: 1	0.05	0.33	0.94	0.48
10–11, NOT_ANY_MORE_NEW, DT: 1	0.05	0.33	0.94	0.48

Table 11 Ensemble of number of communities and central nodes on unweighted network

Type	Chng	Precision	Recall	F1-meas.
3–5, SIZE, DT: 2	0.05	0.62	0.81	0.70
3–5, NEW, DT: 2	0.05	0.62	0.81	0.70
3–5, NOT_ANY_MORE, DT: 2	0.05	0.62	0.81	0.70
3–5, NOT_ANY_MORE_NEW, DT: 2	0.05	0.62	0.81	0.70
9–9, SIZE, DT: 2	0.05	1.00	0.50	0.67
9–9, NEW, DT: 2	0.05	1.00	0.50	0.67
9–9, NOT_ANY_MORE, DT: 2	0.05	1.00	0.50	0.67
9–9, NOT_ANY_MORE_NEW, DT: 2	0.05	1.00	0.50	0.67
32–90, SIZE, DT: 2	0.05	0.58	0.69	0.63
32–90, NEW, DT: 2	0.05	0.58	0.69	0.63
32–90, NOT_ANY_MORE, DT: 2	0.05	0.58	0.69	0.63
32–90, NOT_ANY_MORE_NEW, DT: 2	0.05	0.58	0.69	0.63
12–16, SIZE, DT: 2	0.05	0.44	0.94	0.60
12–16, NEW, DT: 2	0.05	0.44	0.94	0.60
12–16, NOT_ANY_MORE, DT: 2	0.05	0.44	0.94	0.60
12–16, NOT_ANY_MORE_NEW, DT: 2	0.05	0.44	0.94	0.60
17–31, SIZE, DT: 2	0.05	0.45	0.62	0.53
17–31, NEW, DT: 2	0.05	0.45	0.62	0.53
17–31, NOT_ANY_MORE, DT: 2	0.05	0.45	0.62	0.53
17–31, NOT_ANY_MORE_NEW, DT: 2	0.05	0.45	0.62	0.53
6–8, SIZE, DT: 1	0.05	0.35	1.00	0.52
6–8, NEW, DT: 1	0.05	0.35	1.00	0.52
6–8, NOT_ANY_MORE, DT: 1	0.05	0.35	1.00	0.52
6–8, NOT_ANY_MORE_NEW, DT: 1	0.05	0.35	1.00	0.52
32–90, SIZE, DT: 1	0.05	0.35	1.00	0.52
32–90, NEW, DT: 1	0.05	0.35	1.00	0.52
32–90, NOT_ANY_MORE, DT: 1	0.05	0.35	1.00	0.52
32–90, NOT_ANY_MORE_NEW, DT: 1	0.05	0.35	1.00	0.52
17–31, SIZE, DT: 1	0.05	0.34	1.00	0.51
17–31, NEW, DT: 1	0.05	0.34	1.00	0.51
17–31, NOT_ANY_MORE, DT: 1	0.05	0.34	1.00	0.51
17–31, NOT_ANY_MORE_NEW, DT: 1	0.05	0.34	1.00	0.51
12–16, SIZE, DT: 1	0.05	0.33	1.00	0.50
12–16, NEW, DT: 1	0.05	0.33	1.00	0.50
12–16, NOT_ANY_MORE, DT: 1	0.05	0.33	1.00	0.50
12–16, NOT_ANY_MORE_NEW, DT: 1	0.05	0.33	1.00	0.50
3–5, SIZE, DT: 1	0.05	0.33	0.94	0.49
3–5, NEW, DT: 1	0.05	0.33	0.94	0.49
3–5, NOT_ANY_MORE, DT: 1	0.05	0.33	0.94	0.49
3–5, NOT_ANY_MORE_NEW, DT: 1	0.05	0.33	0.94	0.49

Table 12 Data set scalability performance analysis

Graph type	Data set size (# of CDR)	Exec. time (s)
Weighted	100,000	15
Weighted	200,000	27
Weighted	400,000	89
Weighted	800,000	308
Weighted	1,600,000	1231
Unweighted	100,000	12
Unweighted	200,000	27
Unweighted	400,000	91
Unweighted	800,000	296
Unweighted	1,600,000	1260

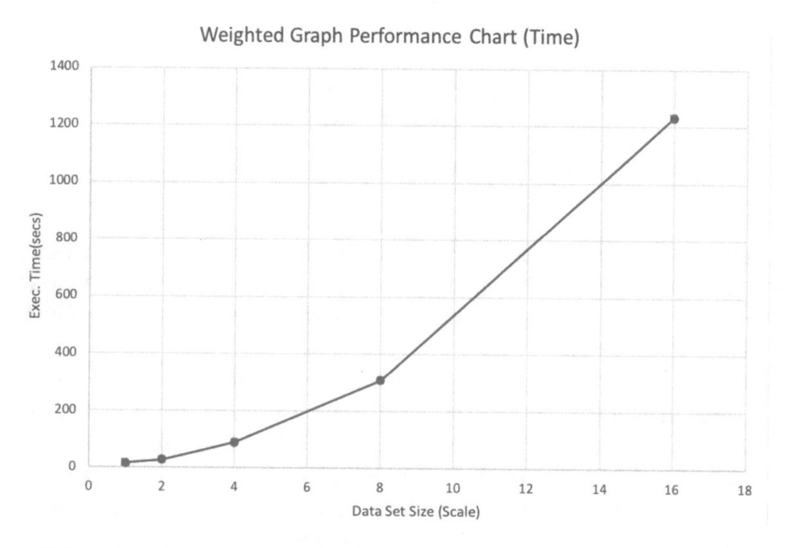

Fig. 2 Weighted graph: exec. time/data set size

few cases, under increasing change threshold, both precision and recall values decrease.

- It is observed that the basic method on tracking the change of central nodes is not sensitive to change threshold. Although high recall values (recall 0.94) are obtained for this method, precision values remain low.
- As in the second method, the method on tracking the change in the community members appears to be insensitive to change threshold. The accuracy values remain the same for each of AVG, MAX and MIN settings. This method provides the highest recall score (recall value 1.00) under AVG and MAX, however, low precision value shows that it has tendency towards labeling time windows as event.
- The accuracy values under weighted and unweighted network structures do not indicate a strong difference. The results are the same for the method with change

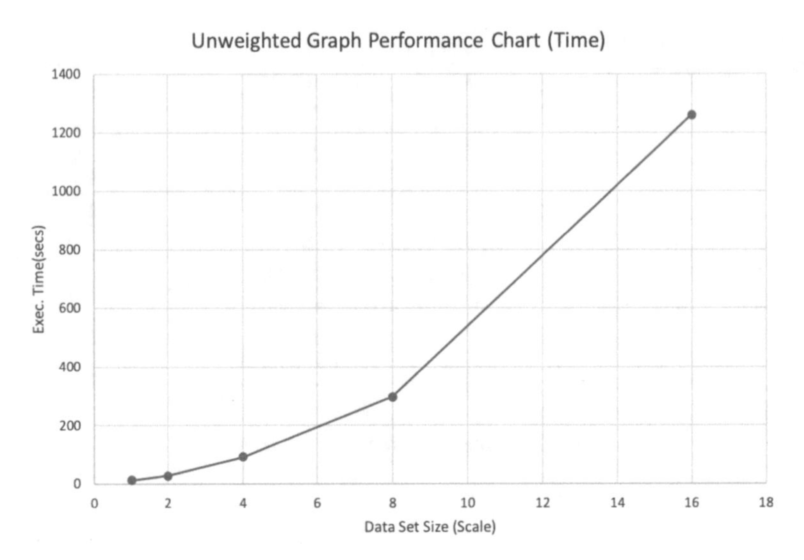

Fig. 3 Unweighted graph: exec. time/data set size

of community members for both of the graphs. For the other two basic and two ensemble methods, the accuracy values slightly higher for the unweighted network structure. Thus, unweighted network structure may be preferable due to its lightweight structure.

- The ensemble method combining the variations of change tracking on number of communities gives the best results by even exceeding the results of number of communities based basic method. The highest F1-measure value is 0.79 for unweighted and 0.75 for weighted directed network. Therefore, we can state that the ensemble method that we combines the results of number of communities change tracking basic method variations is successful.
- On the other hand, the ensemble method that combines the results of number of communities and central nodes based event detection method variations do not give better results than the basic methods. The highest F1-measure value is 0.67 for unweighted and 0.70 for weighted directed network.
- The scalability experiments reflect the quadratic complexity of the employed community detection method in increasing time costs. This time cost is considered to be similar to previous solutions that rely on the change the graph attributes, such as node centrality [18, 19]. However, such previous studies did not report time cost, we can not provide a quantified comparison for scalability analysis.

5 Conclusion and Future Work

Considering that the events may trigger change in the communication pattern of the users, in this work, we model the event detection problem in terms of change detection in community structure within communication network. The proposed approach involves community detection within communication network, and tracking the change in communities along a timeline. We propose event detection under three basic change models: the change in the number of communities, the change in the central nodes of the communities and the change in the size of each community. Additionally, we propose two hybrid change models that are the ensemble of the number of communities change tracking variations and the ensemble of the number of communities and central node change tracking variations. Experiments conducted on a benchmark data set under various settings show that change in the number of communities is a stronger indication for an event.

This work can be extended in several dimensions. As the first one, a learning based method can be devised to determine the optimal change values under various change models as features. As another extension, alternative link based ranking methods can be employed for determining central nodes within communities.

References

1. S. Afra, T. Ozyer, J. Rokne, Netdriller version 2: A powerful social network analysis tool, in *2018 IEEE International Conference on Data Mining Workshops (ICDMW)* (Nov 2018), pp. 1475–1480
2. R. Aktunc, I.H. Toroslu, M. Ozer, H. Davulcu, A dynamic modularity based community detection algorithm for large-scale networks: DSLM, in *Proceedings of the 2015 IEEE/ACM International Conference on Advances in Social Networks Analysis and Mining 2015. ASONAM '15* (2015), pp. 1177–1183
3. R. Aktunc, I. Toroslu, P. Karagoz, Event detection by change tracking on community structure of temporal networks, in *ASONAM* (2018), pp. 928–931. https://doi.org/10.1109/ASONAM.2018.8508325
4. J. Allan, Topic detection and tracking, in *Introduction to Topic Detection and Tracking* (Kluwer Academic Publishers, Norwell, 2002), pp. 1–16. http://dl.acm.org/citation.cfm?id=772260.772262
5. F. Atefeh, W. Khreich, A survey of techniques for event detection in Twitter. Comput. Intell. **31**(1), 132–164 (2015)
6. V. Blondel, J. Guillaume, R. Lambiotte, E. Mech, Fast unfolding of communities in large networks. J. Stat. Mech. 2008, P10008 (2008)
7. Y. Dong, F. Pinelli, Y. Gkoufas, Z. Nabi, F. Calabrese, N.V. Chawla, Inferring unusual crowd events from mobile phone call detail records. CoRR abs/1504.03643 (2015)
8. N. Eagle, A. Pentland, D. Lazer, Inferring friendship network structure by using mobile phone data. Proc. Natl. Acad. Sci. **106**, 15274–15278 (2009)
9. S. Fortunato, Community detection in graphs. Phys. Rep. **486**, 75–174 (2010)
10. Q. Gong, Y. Chen, X. He, Z. Zhuang, T. Wang, H. Huang, X. Wang, X. Fu, Deepscan: exploiting deep learning for malicious account detection in location-based social networks. IEEE Commun. Mag. **56**(11), 21–27 (2018)

11. M. Imran, C. Castillo, F. Diaz, S. Vieweg, Processing social media messages in mass emergency: survey summary, in *Companion Proceedings of the Web Conference 2018* (2018), pp. 507–511
12. I.A. Karatepe, E. Zeydan, Anomaly detection in cellular network data using big data analytics, in *European Wireless 2014; 20th European Wireless Conference* (May 2014), pp. 1–5
13. M.E.J. Newman, Modularity and community structure in networks. Proc. Natl. Acad. Sci. **103**, 8577–8582 (2006)
14. G.K. Orman, V. Labatut, H. Cherifi, Comparative evaluation of community detection algorithms: a topological approach. CoRR abs/1206.4987 (2012). http://dblp.uni-trier.de/db/journals/corr/corr1206.html#abs-1206-4987
15. O. Ozdikis, P. Senkul, H. Oguztuzun, Semantic expansion of tweet contents for enhanced event detection in Twitter, in *International Conference on Advances in Social Networks Analysis and Mining (ASONAM)* (2012), pp. 20–24
16. O. Ozdikis, P. Karagoz, H. Oğuztüzün, Incremental clustering with vector expansion for online event detection in microblogs. Soc. Netw. Anal. Min. **7**(1), 56 (2017)
17. T. Ozyer, R. Alhajj (eds.), *Machine Learning Techniques for Online Social Networks*. Lecture Notes in Social Networks (Springer International Publishing, Cham, 2018)
18. S. Rayana, L. Akoglu, Less is more: building selective anomaly ensembles with application to event detection in temporal graphs, in *SDM* (2015)
19. S. Rayana, L. Akogli, Less is more: building selective anomaly ensembles. ACM Trans. Knowl. Discov. Data **10**(4), 42:1–42:33 (2016)
20. J. Sankaranarayanan, H. Samet, B.E. Teitler, M.D. Lieberman, J. Sperling, TwitterStand: news in tweets, in *ACM SIGSPATIAL International Conference on Advances in Geographic Information Systems (GIS)* (2009), pp. 42–51
21. L. Tang, H. Liu, *Community Detection and Mining in Social Media*. Synthesis Lectures on Data Mining and Knowledge Discovery (Morgan and Claypool Publishers, San Rafael, 2010). https://doi.org/10.2200/S00298ED1V01Y201009DMK003
22. V.A. Traag, A. Browet, F. Calabrese, F. Morlot, Social event detection in massive mobile phone data using probabilistic location inference, in *2011 IEEE Third International Conference on Privacy, Security, Risk and Trust and 2011 IEEE Third International Conference on Social Computing* (Oct 2011), pp. 625–628
23. L. Waltman, N.J. van Eck, A smart local moving algorithm for large-scale modularity-based community detection. CoRR abs/1308.6604 (2013). http://dblp.uni-trier.de/db/journals/corr/corr1308.html#WaltmanE13
24. M. Yu, Q. Huang, H. Qin, C. Scheele, C. Yang, Deep learning for real-time social media text classification for situation awareness – using hurricanes Sandy, Harvey, and Irma as case studies. Int. J. Digital Earth **0**(0), 1–18 (2019)
25. Z. Zhang, Q. He, J. Gao, M. Ni, A deep learning approach for detecting traffic accidents from social media data. Transp. Res C Emerg. Technol. **86**, 580–596 (2018)
26. X. Zhou, L. Chen, Event detection over Twitter social media streams. VLDB J. **23**(3), 381–400 (2014)

Chasing Undetected Fraud Processes with Deep Probabilistic Networks

Christophe Thovex

1 Introduction

Intelligence is not what we know, but what we do when we do not know. This statement is commonly attributed to the well-known psychologist JEAN PIAGET, biologist and epistemologist who agreed with the importance of communication and social interactions for the development of language and of human intelligence, as the philosopher Putnam [1]. According to the historical experimentation supposed to have been lead by the roman emperor FREDERICK II HOHENSTAUFEN (1197–1250), when babies cannot communicate with other babies and adults, after a few years they have no cognitive ability to learn language then they cannot acquire normal intellectual capabilities. Hence, human learning is social, and social learning—the action of learning aided by social networks (not necessarily digitized networks)—is essential to foster or accelerate the development of intelligence [2]. This forms a first statement.

Nevertheless, according to [3], babies start to communicate by signs before being physically able to talk and verbalize thoughts with adults. Babies and animals such as dolphins and primates fail to the TURING's test, an imitation game designed to proof the existence of thought in computing machinery [4]. Obviously,

C. Thovex (✉)
DATA2B, Cesson Sevigné, France

Laboratoire Franco-Mexicain d'Informatique et d'Automatique, UMI CNRS 3175 – CINVESTAV, Zacatenco, Mexico D.F., Mexico
e-mail: christophe.thovex@data2b.net
https://data2b.net

M. Kaya et al. (eds.), *Putting Social Media and Networking Data in Practice for Education, Planning, Prediction and Recommendation*, Lecture Notes in Social Networks, https://doi.org/10.1007/978-3-030-33698-1_6

the TURING's test fails to detect thought but estimates the ability of computing machinery to interact with natural language skills. It shows how much *computing machinery only performs what we know how to order it to perform*, as stated by ADA OF LOVELACE [4]. This forms a second statement.

Unifying the first and second statements, we may accept the eventuality of an intelligence to be developed with computing machinery without interpreted and outer evidence of its existence—i.e., absence of evidence is not evidence of absence. However, in 2018 we were still unable to say a machine how to play the imitation game and handle the TURING's test more than a few minutes, while adult humans cannot fail to the test, at least in their native language.[1] Anyway, IT reminds us how fair was the conclusion of A. TURING about thinking machines: "We can only see a short distance ahead, but we can see plenty there that needs to be done" [4].

Since precursor works such as presented in [5], the hypothesis of a self-programming machine offers the promise of an Artificial Intelligence (AI), *performing what we do not know how to order it to perform*. PITRAT and its general problems solver CAIA opened the Pandora box providing mathematical solutions to some specialized problems that were never coded, thanks to meta-knowledge [6]. Because the machine still does not develop itself but does nothing without human programming, it entails there could be no AI in machine learning, just *artificial knowledge* finally. Talking machines (*i.e.*, chatbots) process patterns in textual representations for performing operations on natural language. Computer vision processes patterns in images, video or 3D representations for objects classification. Both systems can collaborate but chatbots "ignore" everything of what computer vision "sees" and conversely. This is where the first and second statements are recalled for theoretical opening—*i.e.*, (1) human learning is social and (2) machine only performs what we know how to order it to perform.

Indeed, human intelligence feeds knowledge forward and knowledge feeds human intelligence backward. Conversely, propagation and back-propagation processes defined in Convolutional Neural Networks (CNN) inherit from human intelligence and feed forward artificial knowledge to knowledge/data, in deep learning models [7–9]. They do not feed backward any intelligence, because the intelligence feeding machine is human and there is no integrated body feeding backward human intelligence with artificial knowledge, only hardware interfaces adapted to human perception. Brain implants of silicon chips would not change this, they could augment senses and perception but would remain separate components still not forming a single body with the grey matter, organic intelligence and brain plasticity [10]. Nevertheless, the diversity of models offered by deep networks structures and neuronal plasticity remain largely unexplored in computer science. This, could be a reason why we still satisfy uncertain problems such as weak signal detection and social opinion mining with exploratory models aiding human experts to decision-making, and probabilistic machine learning [11, 12]. There were the start ideas of the presented work, which aims at learning structural

[1] Human fails at the Turing's test invalidate the test.

knowledge from big data and social context as human intelligence does for social learning, hoping to foster or accelerate the development of artificial knowledge—*i.e.*, *Analytic Intelligence* so as to mislead nobody about the nature and results of analytic algorithms and brute force. Fraud detection is a classical example of complex task requiring human expertise, with the risk of errors that it also represents when processing large and complex data. The *Facsimile* case represents a recurrent problem concerning fraud detection. There is apparently no formalization of such a case as a problem or a class of problems. A proper definition is introduced prior to the presented work.

A *facsimile* is a conform copy of a document, literally. Extending this notion to any kind of support and information enables to define a class of issues in fraud detection, the *facsimile* problem. It concerns all cases of successful frauds based on the conformity of the fraud process with the normal process, turning both processes statistically similar and so undetectable by common models in machine learning. As an example, a fake banknote that would be perfectly conform to a genuine banknote with the same identification code (*i.e.*, a facsimile) would not allow to say which one is official and which one is fake based on their physical comparison. Thus, making such a fake banknote would ensure a 100% successful fraud process –an infinite loss potential–, given the unique dataset that would represent the physical characteristics of both banknotes—*i.e.*, endogenous data. According to [13], just before the year 2000 a reduction of 2.5% of credit cards frauds triggered a saving of one million dollars a year.

To detect a *facsimile* fraud process would necessarily require exogenous data—*i.e.*, data tied to the banknotes and external to their intrinsic form—such as owners history or geographical tracks forming distinctive datasets, respectively to each banknote. Without exogenous data, facsimile issues remain insolvable. In real life, exogenous data are not often available with endogenous data and fakes are never perfect, but can be so close to the original that they mislead experts and detection algorithms as facsimile could do. Consequently, tackling some of the most efficient fraud processes depends on solutions to facsimile problems.

This work was first presented as a short paper at the IEEE/ACM international conference Advances in Social Network Analysis and Mining, ASONAM 2018, Barcelona, Spain [14]. Providing various additional elements about the modeling process, this chapter largely extends the first publication. It introduces and describes a deep network structure of supervised learning for the detection of fraud process close to facsimile in business data. It was first aiming at facing a pseudo-facsimile problem encountered by a French public institution (confidential). Due to unexpected results observed in certain circumstances during experimentation, the model and its theoretical foundations were more carefully studied and the protocol and results were checked twice, in order to avoid any error in their communication. All sections were improved in this chapter and a particular attention was paid to the model presentation, with numerous details helping serious readers to understand and evaluate its relevance for various case studies.

Section 2 presents previous works providing theoretical foundations and technical bases to the presented work. Section 3 proposes a structural and probabilistic alternative of deep network inspired from the time reversal phenomena, for signal interference in finite spaces as a solution to the facsimile problem. Section 4 exposes an offline experimentation of the proposed model based on a large dataset built from business process databases, including training and test datasets. Interactions between people and/or actions/items stored in the data define the topology of a social network. Section 5 presents the outcomes and additional observations resulting from the experimentation. Lastly, Sect. 6 discusses the conclusions and applications perspectives.

2 Related Works

Artificial neural networks are hierarchical networks inspired from the observation of biological neurons solicited for vision [7, 15]. Defined as Graph Transformer Networks (GTN), they allow supervised machine learning from a sample of labeled or selected data (texts, pictures, audio), thanks to back-propagation process and are able to minimize the output error rate in a multi-layers comparative process of samples and sub-samples. In [7], Convolutional Neural Networks (CNN) combine an optimizing method such as gradient descent with sub-sampling layers to find the optimal response path for an input, regarding learned features and parameters. For pictures or handwriting recognition, the analytic redundancy produced by deep multi-scales layers (up to 20 layers) organized in CNN provide lower error rates and high performances with low memory space, compared with models integrating individual and multiple modules, according to [7, 9]. GTN and CNN remind us the efficiency of biomimetics in computer vision [7, 15], then in other applications of computer science requiring pattern-recognition such as fault detection in power systems [16].

Neuronal structures in biological bodies are not only processing vision, pattern recognition and similarity-based classification problems such as do CNN in computer vision. Therefore, dealing with rules-discovery or semantics for problems solving—i.e, machine reasoning—could require different heuristics, neuronal processes and structures than those usually found in deep learning. As an example, Generative Adversarial Networks (GAN) seek for a MinMax solution in a zero-sum game while CNN processes to parallel optimization [17, 18]. Reinforcement learning also extracts new application perspectives from the game theory, enabling the dynamic optimization or retrieval of intrinsic causes from small groups interactions [8]. Based on the biomimetic metaphor, many more different approaches of neural networks might be defined, in AI such as in social networks analysis and mining, for other problems than parallel optimization. Furthermore, some inferential approaches in knowledge management, constraints satisfaction problems or other research domains provide significant results without biomimetic approach [6].

Inverse problems enable to retrieve original state from a derived state in a data space—*e.g.* picture, signal. In [19], wavelets processing as an inverse problem enhances images from large astronomical instruments. Solutions are generally based on statistics or probabilistic estimators or optimizers such as least squares or Bayesian models [8, 20, 21]. Due to chaotic phenomena, retrieving the accurate location of an original signal in a complex and finite space is not satisfied with enough accuracy and confidence as an inverse problem but in a wave-defined space, the specific approach defined as time-reversal enables to locate and reproduce the original signal with reliable results [22]. It presents a particular robustness to complex data systems and uncertainty for retrieving the original characteristics of weak and/or complex signals in finite data spaces. Such a property might be propitious to retrieve and locate undefined features conjunctions within deep and/or social networks structures, regarding the *facsimile* problem.

Graph models remain theoretical tools relevant in numerous domains. They are unavoidable in knowledge and information retrieval from/for the World Wide Web [23, 24]. Since the introduction of sociograms in social psychology by Moreno [25], social networks analysis and mining is mostly based on graph models [26–30]. Artificial neural networks present obvious similarities, but also divergences with Markovian networks detailed in [31]. Based on [32], Bayesian networks produce interesting results in fraud detection but are not appropriate for statistically unsignificant data characterizing the facsimile problem. Probabilistic propagations in social networks for risk analysis is also introduced in [33]. In order to keep track with the rapid changes of organized attacks and patterns of credit card fraud, automatic learning fraud models are needed instead of predefined rules-based expert systems [13]. Time reversal in wave-defined spaces meets this requirement [22]. However, in large and/or deep networks complexity and computing costs induced by Markov chains with memory entail a fine tuning or alternative models alleviating complex processes such as stochastic propagation, thanks to abstract simplifications [34]. In [17], paired neural networks with antagonist rules avoid the Markov chains for training, using exact propagation instead of stochastic approximation.

As noticed in [7], most of the deep networks structures for machine learning might be attached to a single class of energy models. The models presented above also belong to Network Science, a multidisciplinary branch of Data Science defined since 2005 by the United States National Research Council as "the study of network representations of physical, biological, and social phenomena leading to predictive models of these phenomena" [35].

Unfortunately, the multidisciplinary foundations of the model presented hereafter are dense and hard to conceptualize at a glance. Its theoretical foundations stem on physical phenomena that are still partially explained, as well as on massive probabilistic interpolations to handle for which computers are much more efficient than people. As a consequence, the author takes care of sharing detailed observations and explanations in the following section, so as to help interested readers in building their proper understandings of the presented work. Furthermore and unfortunately, the large data set having originally motivated that work and employed for experimentation must definitely remain confidential, for efficiency reasons.

3 Resonance and Interference as a Probabilistic Network

In [13], the proposal shorten raw association rules learned from symbolic data (transactions data, 38 fields, 11,700 rows) into generalizations for detecting misuses with neural data mining. A strong interaction is observed in the results between confidence/precision and fraud discovery/recall. Even combining a credit network with a time network, raising recall results in the heavy reduction of precision, down to 1%. A trade-off of about 80% of confidence for about 100% recall is finally obtained but necessitates two different rules-based expert systems for training, which is not compliant with the initial concerns about the rapid changes of organized attacks and patterns.

Based on current observations, the presented proposal results from a multidisciplinary study leading to a disruptive approach defining a model of Deep Resonance Interference Network (DRIN). It is designed for big data (thousands of columns and billions of rows), but convenient for smaller data sets. It aims at providing a solution for facsimile problems and fraud detection that is not affected from the rapid changes of organized attacks and patterns (rules-free system) and let the experts focus on visual frauds cases and/or risks, instead of chasing unknown and local rules in perpetual change, which are sometimes in conflict at global scale.

In large data sets, the variety of variables and values induces the loss of subtle local characteristics in the mass of information. It entails a phenomenon that presents a conceptual similarity with the chaotic phenomenon tackled thanks to the wave-based model for time reversal defined in [22]. The time reversal model also presents an interesting conceptual similarity with CNN as both approaches are based on feed-forward and back-propagation processes. Basically, these conceptual observations supply the foundations of an hypothesis by which appears a possible solution for the detection of fraud cases related to facsimile problems.

Learning and discriminating features from stochastic gradient descent, CNNs are described as optimizers in [17]. Deep Resonance Interference Network is not a large combinatorial optimizer but a *correlator*. Its purpose is not to classify an input based on trained features, but to correlate unknown features based on training instances of a unique feature. Such as with the separated processes presented in [34], DRIN enables to mine huge data spaces as separated data sub-spaces and to retrieve undefined features conjunctions as well within as across separated sub-spaces. As for GAN, it avoids stochastic approximation for training large data sets in affordable execution costs in time and space [17].

DRIN first defines a trivial function to propagate exact values in a multidimensional way instead of a hierarchical way, through the multiple layers as a training signal to be compared with a discrete snapshot of a pure wave in the data space. The training values are based on a certified training set and a discrete signal representation is stored in nodes and arcs. Then the propagation process pushes the signal in a hierarchical way as a probabilistic relationship, simulating interferences and resonances across the whole data space. Interferences and resonances represent an energy model emanating through nodes within a dense network structure. Local divergences in-between nodes and flows resulting from the two-phases propagation process within the simulated waves field are smoothed as average probabilities, trivially making the model globally coherent with the energy propagation rules in networks—*e.g.*, node and mesh method [36]. Then DRIN triggers a back-propagation process strictly reversing the hierarchical propagation process through the coherent global state obtained, resulting in a second and globally coherent state. Back-propagation generates here a coherent amplification of the variance in interferences and resonances within the discrete wave field simulated, avoiding any global energy loss or amplification along the probabilistic paths of the deep network structure. As a result, unknown features and features conjunctions correlated with the instances of the training set T are expressed in probabilistic values such as $p \in [0; 1]$, unveiling the risk of facsimile fraud case within the whole data space and sub-spaces, with T set of certified instances of a known fraud feature.

The basic design of the model is formalized for a large social dataset in Sect. 3.1. The implementation of DRIN core concept is then detailed in Sect. 3.2. It was implemented and experimented as a solution to the *facsimile* problem. However, as a combinatorial correlator it should find other applications in analytic intelligence, might be for machine reasoning with epistemic properties and inferences such as introduced in [37, 38] or [39].

3.1 Bases of a DRIN

Let v a variable in a big data set $\{DS\}$ such as $\{DS\}$ is a superset of m subsets $\{D\}$ and each $\{D\}$ is a subset of relations $r(v_k, v_{k+1})$ in which all (v_k, v_{k+1}) are unique in $\{D\}$ and in $\{DS\}$.[2] There exists a subset $\{S\}$ of sequences s such as $s\big(r_{k=1}(v_k, v_{k+1}), r_{k+1}(v_k, v_{k+1}), \ldots, r_n(v_{n-1}, v_n)\big)$ with $1 < n < +\infty$, in $\{DS\}$. This defines a NOSQL[3] and generic scheme of database for big data applications—*i.e.*, a nondescript relational scheme for large heterogeneous databases without normalization and integrity constraints.

Let $\{DS\}$ represent a complex business process such as found in trading, client relationship management, companies accountancy etc. Let v be the root variable on which we focus for frauds detection related to facsimile cases—*e.g.*, v might represent products, categories, people, accounts, quantities, modalities etc. To facilitate understanding, the root variable v might be compared with the root value in the Erdos-Renyi algorithm [40].

Let a sequence $uds\big(r_{k=1}(v_k, v_{k+1}), r_{k+1}(v_k, v_{k+1}), \ldots, r_n(v_n, v_{n+1})\big)$ in $\{DS\}$. It is user-defined such as $v = v_k$ in $r_{k=1}(v_k, v_{k+1})$ of *uds*—*i.e.*, the root variable is at the first rank of *uds*. The length of *uds* enables users to control the trade-off

[2]Each $\{D\}$ may be thought as a facts table in $\{DS\}$.

[3]Acronym for Not Only SQL, databases without integrity constraints.

between computing costs and results completeness/relevance. With uds including all relations $r(v1, v2)$ in $\{D\}$, computing costs are maximal for $\{DS\}$.

For each user-defined sequence uds, DRIN builds a highly connected networks, persistent or in-memory depending on the chosen implementation, and returns the set of bipartite graphs represented by uds, fully valued with probabilistic weights resulting from a feed-forward and back-propagation process detailed later in this section. For instance, with $uds = \big(r_1(v_1, v_2), r_2(v_2, v_3)\big)$ the directed and virtually symmetric graph returned is valued and persistent. It is defined as $OG(N, A)$, with (N, A) respectively sets of nodes and of arcs such as:

$$N : \forall \quad v_x \in v_1 \cup v_2 \cup v_3 \quad \exists \quad n \in N : n = v_x, v = v_1 \tag{1}$$

$$A : \forall \quad a(n, n') \in A \quad \text{with} \quad n \in N \wedge n' \in N$$

$$\exists \quad (n, n') : n = v_1 \wedge n' = v_2 \wedge \exists\, r_1(n, n')$$

$$\vee \quad (n, n') : n = v_2 \wedge n' = v_3 \wedge \exists\, r_2(n, n')$$

$$\vee \quad \exists \quad va(n', n) \in A$$

Equation (1) defines a complete virtually symmetric network $G(N, A)$ representing all possible sequences $s = (v1, v2, v3)$ such as the root variable v is at the first rank and there virtually exists $vs = (v3, v2, v1)$. Virtual arcs $va(n', n)$ are not added in A as they are temporarily processed as abstract objects for computing enhanced variances during the back-propagation process, such as explained beforehand.

3.2 Core of a DRIN

The core concept of DRIN inherits from the physical modeling of multidimensional waves interference in finite space, developed for time reversal applications and presented in [22]. The basic design of DRIN projects this paradigm within a multidimensional and deep network structure, appearing as a promising approach for tackling problems such as *facsimile* and undetected fraud cases in big data sets. Related works on Markov chains with graphs, focused on mixed memory models and on conceptual similarities and differences between neural and Markovian networks, compared with antagonist models avoiding stochastic approximations appeared helpful for converging towards such a multidisciplinary approach [17, 31, 33, 34].

DRIN defines a probabilistic feed-forward and back-propagation process, inspired from the quotations above and mimicking time reversal within deep heterogeneous networks. It covers the finite and discrete space of the superset $\{DS\}$ with a deep heterogeneous graph, then implements a mixed form of Bayesian network for the artificial propagation of probability values from a uniform training

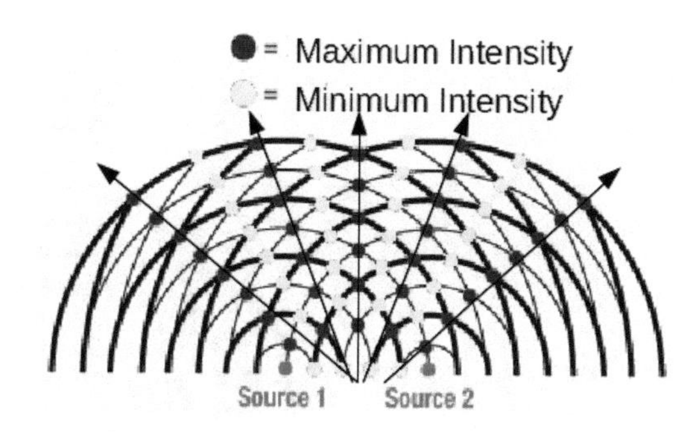

Fig. 1 Two sources physical interference pattern

set. As a result, it makes appear the local amplification/absorption of conjugated probabilities as a signal for fraud risks, mapped onto the nodes and arcs of the output network $OG(N, A)$. In Fig. 1, physical interference pattern with two sources is presented for illustration.[4] Resonances are symbolized by (red) dark dots and interferences by (blue) light dots (phase cancellation). While Fig. 1 illustrates the combinatorial width of a two-sources and two-dimensions data space (wave amplitude being a hidden dimension), one must imagine the phenomenon simulated with DRIN in a big data space of hundreds or thousands of columns and millions of rows or more, each root value in the training set behaving like a source. Given a dataset $\{DS\}$, a root variable v and a user-defined sequence uds such as previously defined, DRIN builds an intermediate graph $IG(N, A)$ covering the sub-space of $\{DS\}$ constrained with uds.

Let $\{T\}$ be the training dataset for $IG(N, A)$ with $\{T\}$ representing the same object class than v—*e.g.*, company, place, date, client, product name, purchased quantity etc. Given $\{V\}$ the set of modalities/values of v, $\{T\} \cap \{V\} \neq \emptyset$.

At the initial state of the DRIN process, $IG(N, A)$ is an open network with $\{N\} = \{V\}$ and $\{A\} = \emptyset$. An element of $\{N\}$ have no proper weight value.

The Propagation Phase The first phase of the DRIN process consists in populating $rc(r_{k=1}, r_{k=2}, \ldots, r_k)$ with valued elements, depending on uds. Given the relation $r_{k=1}(v_k, v_{k+1})$ in uds,

$$\forall\, v_k : \exists\, v_k \in \{T\}, P(v_k) = 1 \tag{2}$$

$$P\big(r_{k=1}(v_k, v_{k+1})\big) = P(v_k)$$

In step 1, DRIN initializes the values of all pairs of variables (v, v_k), with a relational probability value $P(v_k)$, assuming the root value $P(v) = 1$ for the certified positive found within the training dataset. An illustration of the relational probability value at Step 1 of DRIN is presented in Fig. 2 for better understanding.

[4]Source: https://astarmathsandphysics.com/ib-physics-notes/127-waves-and-oscillations/1492-the-two-source-interference-pattern.html.

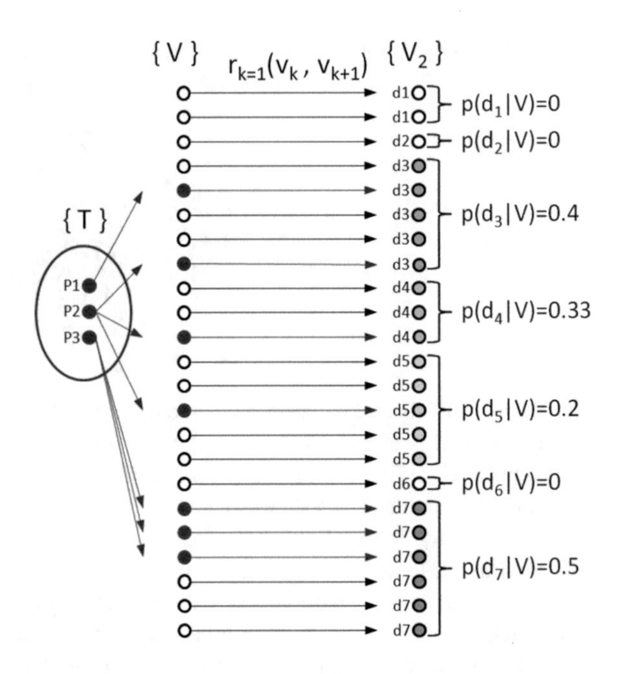

Fig. 2 Relational probability value during Step 1—DRIN

It means the root variable in uds takes $P(v) = 1$ when it exists into T and $P(v_k)$ is the probability of a v positive for each v_k within the relation (v, v_k), with $P(v, v_k) = P(v)/|(v, v_k)|$.

The root variable v_0 (assuming $v = v_0$) simulates a source diffusing a signal into a discrete plane. Then the process propagates the root probability v_0 from $v_{k=1}$ towards v_n within the arcs $a(v_k, v_{k+1})$ representing all relation/bipartite graphs $r_{k=1}(v_k, v_{k+1})$ in uds.

Given the relations chain $rc(r_{k=1}, r_{k=2}, \ldots, r_k)$, with $r_k(v_k, v_{k+1})$ and $k > 1$, the initial value of $P(v_k)$ is calculated from $rc(r_{k=1}, r_{k=2}, \ldots, r_k)$ and the initial value of $P(v_{k+1})$ is calculated from $rc(r_{k=1}, r_{k=2}, \ldots, r_{k+1})$. In such a relations chain rc, there is no continuous orthogonality from r_k to r_{k+1}. As a consequence rc cannot represent a continuous and discrete plane but a set of distinct and discrete planes coherent one-to-one. It means that rc represents a single multidimensional and discrete space.

Let $f(n_k, p)$ be a function returning the initial probability of a node n_k from $IG(N, A)$ in $r_k(v_k, v_{k+1})$, such as $f(v_k, 1)$ calculates

$$\frac{|rc(r_{k=1}, r_{k=2}, \ldots, r_k) : P\big(v_k \in r_{k=1}(v_k, v_{k+1})\big) = p = 1|}{|rc(r_{k=1}, r_{k=2}, \ldots, r_k)|} \tag{3}$$

$f(n_k, p)$ returns the proportion of true positive of same object nature than the root variable in indirect relation with any variable across the relations chain $rc(r_{k=1}, r_{k=2}, \ldots, r_k)$. This value is named initial risk index, $niri(n_k)$. It simulates the punctual sample of a wave field resulting from the propagated signal (training set) within the represented space.

In step 2, DRIN sets all nodes n of $IG(N, A) : n \notin \{V\}$ (not root nodes) with an initial risk index $niri(n_k)$ thanks to $f(n_k)$. It ensures a trivial and reliable state for the following steps.

In step 3, DRIN sets the arcs of $IG(N, A) : \forall\ a(n, n') \in A\ n \notin \{V\}$ with an initial index named mean probability of propagation (mpp). This index is based on the initial risk index and incoming chains of n. Given $r_{k>1}(v_k, v_{k+1})$, it is defined as follows.

$$mpp\big(r_{k>1}\big) = \left(\frac{\sum P\big(r_{k-1}(v_k, v_{k+1})\big)}{|r_{k-1}(v_k, v_{k+1}|} + niri(v_k)\right)/2 \tag{4}$$

Equation (4) defines the mean probability from $P(v_k, v_{k+1})|P(v_{k-1}, v_k)$ and $niri(v_k)$. It propagates the incoming signals from a variable of order $k - 1$ towards a variable of order $k + 1$ depending on the relations chains of order k and $k + 1$. The signal is smoothed by the transition variable of order k (Eq. (4)) and propagated within the entire space of uds (user-defined sequence), or in the entire space of $\{DS\}$ when uds includes all the relations $r(v1, v2)$ of all subsets $\{D\}$. It results in a complete network simulating a discrete wave field and local resonances or interferences, within the represented space.

In step 4, DRIN replaces the initial risk index, having ensured a reliable state for the first steps, with a transition probability. The goal is to avoid punctual incoherence in nodes within the simulated wave field before to process it with back-propagation. It is achieved thanks to a couple of new values named local transition probability (ltp) and conditional propagation probability (cpp). Local transition probability represents a local probability of signal transmission for a node $n_k : k > 1$, based on its local context known thanks to the previous steps. The parametric function $f(n_k, p)$ (Eq. (3)) is already defined for that, in a way that when $p = 0$ it returns

$$P\big(v_k \in r_{k=1}(v_k, v_{k+1})\big) > 0$$

instead of

$$P\big(v_k \in r_{k=1}(v_k, v_{k+1})\big) = 1$$

So $f(v_k, 0)$ returns the proportion of positive elements of $\{V\}$ in relation to v_k across the relations chain $rc(r_{k=1}, r_{k=2}, \ldots, r_k)$ (with $r_k = r_k(v_k, v_{k+1})$). In case there exists no v_{k+1} it does not fail, but for improved reading it is trivially defined as

$$\frac{|rc(r_{k=1}, r_{k=2}, \ldots, r_k) : P\big(v_k \in r_{k=1}(v_k, v_{k+1})\big) > 0|}{|rc(r_{k=1}, r_{k=2}, \ldots, r_k)|} \tag{5}$$

Conditional propagation probability is trivially defined as

$$cpp(v_k) = \big(niri(v_k) + f(v_k, 0)\big)/2 \tag{6}$$

Equation (6) smoothes propagation chains for alleviating the bias that results from the initial risk index of nodes n_k, needed as a bootstrap in DRIN.

It is important to notice that the relations chain $rc(r_{k=1}, r_{k=2}, \ldots, r_k)$ does not lose path information in DRIN. By the way, rc stores full end-to-end paths for each chain ($v_k = 1, \ldots, v_k - 1, v_k$), so that there can be several $n \in \{N\}$ having the same label (value) in $IG(N, A)$. This particularity enables to conserve a maximal probability density from $f(v_k, 0)$, in the space of representation exposed by usd as in $\{DS\}$. Such as with time reversal and wave-based models able to deal with chaotic phenomena, it is essential for avoiding the degradation of original signal, being amplification or energy loss.

At the end of the propagation phase, $IG(N, A)$ still comprises all $n : n \in \{T\}$ and the relations chains starting from these nodes. It means $\{V\}$ the set of root values still contains at least one element of the training dataset $\{T\}$. This is obviously useless when the purpose is to detect fraud cases and we later exclude all training elements from $IG(N, A)$ to ensure its relevance. DRIN offers the choice to delete all elements remaining from the training dataset in the output network $OG(N, A)$, in order to focus on the detection of new fraud cases without disturbing the estimations of recall and precision for experimentation (*f measure*). Nevertheless, for certain use cases they remain useful to the domain experts for the exploratory analysis, visualization and understanding of endogenous data related to facsimile problems.

The Back-Propagation Phase The propagation phase of DRIN simulates a discrete wave field and resonances or interferences from a training signal in a multidimensional and coherent space. The back-propagation phase simulates a reversal as an echo of the training signal within the wave field, for a better separation of resonances and interferences in final representations resulting from the model.

The multidisciplinary heuristic that DRIN aims at implementing is mostly defined in the propagation phase, which can be thought as a probabilistic signals propagation in a finite and discrete space without degradation of spatial distribution and of the signal-to-noise ratio entailed by the simulated wave field. Therefore, in order not to introduce bias in $IG(N, A)$, which state at the end of propagation phase is coherent, the back-propagation process must faithfully reverse the propagation process. However, the bootstrap step (step 1 of DRIN) may not be avoided for a regular interference generation before to focus the signals reversal onto the chosen root variable. As a result, the back-propagation process in DRIN is fast to define, even though its execution remains greedy.

In step 5, DRIN returns the signals represented by risk probability values across the relations chains exposed by usd, from r_k to $r_{k=1}$. Based on Eq. (4), the reversal of mean probability of propagation is defined in Eq. (7). In case there exists no v_{k+1} it does not fail, but for improved reading it is trivially defined as

$$rmpp(r_k) = \left(\frac{\sum P(r_k(v_k, v_{k+1}))}{|r_k(v_k, v_{k+1}|} + cpp(v_k) \right) /2 \tag{7}$$

For better understanding, it is reminded that $k + 1$ is the highest order value in uds when starting the back-propagation process. Equation (7) defines the reversed mean probability from $P(v_{k-1}, v_k)|P(v_k, v_{k+1})$ and $cpp(v_k)$. It propagates the incoming signals from a variable of order $k + 1$ towards a variable of order $k - 1$ depending on the relations chains of order k and $k - 1$.

As with Eq. (5), a reversal function $rf(n_k, p)$ is defined as follows

$$\frac{|rc(r_k, r_{k-1}, \ldots, r_{k=1}) : P(v_k \in r_k(v_{k-1}, v_k) < 1)|}{|rc(r_k, r_{k-1}, \ldots, r_{k=1})|} \tag{8}$$

Equation (8) slightly improves the dispersion of probability values, due to the fact it excludes all maximal values $p(v_k) = 1$ in rc from the final state of $IG(N, A)$.

In facsimile problem, an item and its facsimile have no quantitative/qualitative differences. As a consequence, there is no interest at discerning their properties and all items of the same type can be considered as atoms. In such a case, information related to each item can be thought as endogenous information, *i.e.*, a probable and unique way to discern original items from facsimiles, as previously stated (Sect. 1). These theoretical considerations are fully assumed in DRIN, which projects a physical model into a big data set and deep learning network for detecting fraud cases related to facsimile problems. Experiments and results are presented in the following sections.

4 Experimentation with Large Data Sets

Before to return an output graph $OG(N, A)$, DRIN builds $IG(N, A)$, a probabilistic and discrete representation of a finite space of n dimensions, with $n = max(k)$ from the user-defined sequence uds. Each different uds returns a new and persistent $OG(N, A)$. As a result, $OG(N, A)$ is the local network related to uds as a local exposition of $\{DS\}$. Due to the way DRIN defines social networks from large heterogeneous data sets without requiring direct people-to-people relationship stored into $\{DS\}$, when at least one variable v_k represents people in uds, $OG(N, A)$ may represent a hidden social network.

Conversely with the directed and symmetric graphs $IG(N, A)$, in $OG(N, A)$ there exists only one node per distinct label (value) from $\{DS\}$ and one arc per distinct pair of nodes (n, n'). It enables to merge different stored $OG(N, A)$ into a single global network so as to provide an overview of several concomitant spaces and to explore them as distinct communities, thanks to social networks analysis methods and tools [26–28, 30]. An example of such a multidimensional network $CG(N, A)$ visualized for exploratory analysis is illustrated in Fig. 3.

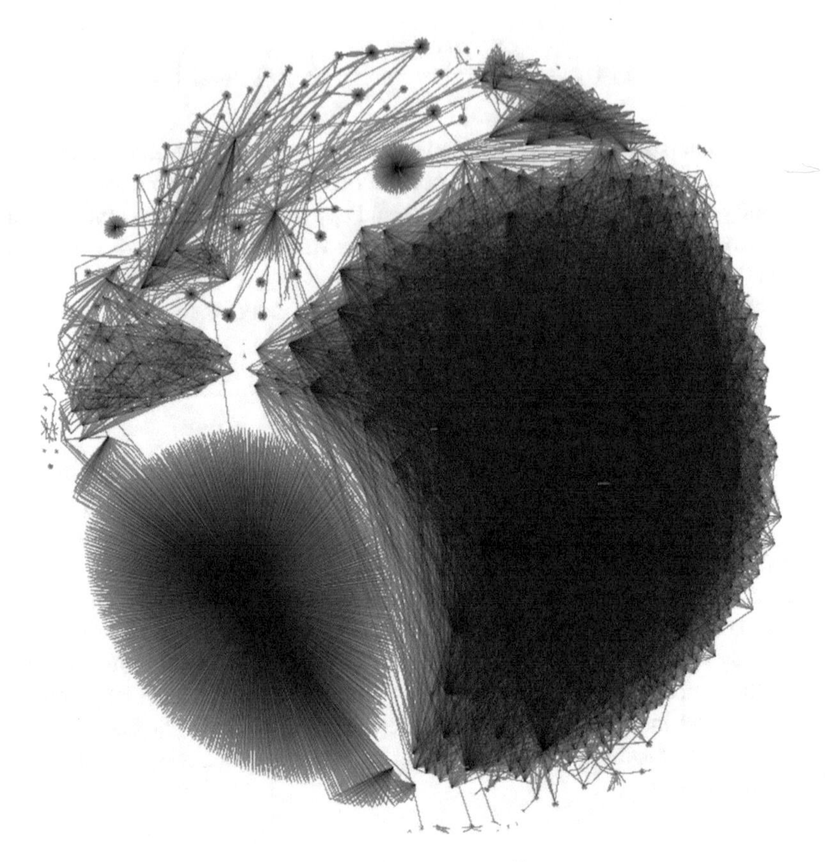

Fig. 3 Composite network merging several $OG(N, A)$—DRIN

The composite network $CG(N, A)$ illustrated (Fig. 3) merges six $OG(N, A)$ of 2 to 19 dimensions. It enables to mine deep hidden networks on small hardware clusters, instead of defining a single $OG(N, A)$ from a maximal user-defined sequence uds covering $\{DS\}$, which could require an important computing cost. An illustration of hidden network structure found by filtering $CG(N, A)$ thanks to its DRIN values is presented in Fig. 4.[5]

DRIN works after-the-fact (offline) with previously stored data sets (batch layer in Lambda architecture). However, large in-memory graph models being propitious to network-computing and parallel processing, it appears relevant to take advantage of DRIN offline processing in order to trigger real-time alerts from an online process standing on the theoretical basements provided (speed layer in Lambda/Kappa architecture).

[5]Figures 3 and 4 were generated from Gephi (v 0.8.2), open source software for graphs analysis—https://gephi.org/.

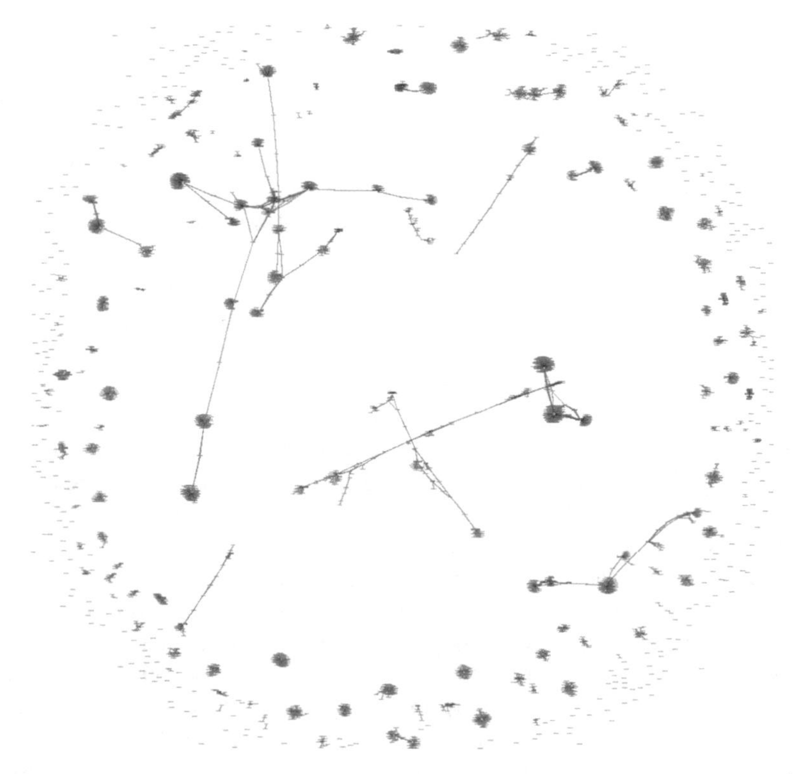

Fig. 4 Mining hidden networks by filtering $CG(N, A)$—DRIN

DRIN was experimented with several databases from about 1 to 100 GB, including data sets up to 1000 variables mapped onto one billion rows. Temporary data sets of about five billion rows were observed with *uds* up to 34 dimensions, from a single 8 cores-16 GB node. Various experimentations with data sets up to 30 TB and a 176 cores-1 TB hardware cluster are under preparation.

5 Outcomes and Observations

Deep learning models generally require large training sets before to ensure satisfying recall and precision measures. In [13], the training dataset is more than half the test dataset, for about 100% recall and 80% precision. As an example, the CNN LeNet-5 requires about 20 passes on a training set of 60,000 patterns before to reach a stable performance [7]. These references are not necessarily to be compared with DRIN as purposes, paradigms and contexts are different for each model, even though they present certain conceptual similarities. Furthermore, the subjacent heuristic to DRIN is a wave-based inverse problem, which might produce different reactions than commonly observed with usual feed-forward and back-propagation network-based model for classification.

DRIN offered two thresholds parameters for filtering the suspicious conjunctions of variables and values (candidate subnets) in its first version, so as to allow users to act on the level of recall and precision. There were no automatic setting of the best parameters. Next versions should fulfill the lack, however empirical tuning is easy and quick for common uses with small or middle-size hardware—*e.g.*, 80 GB-16 cores.

DRIN is a recurrent learning model which can improve its performance adding new positives in the training set from previous results [8]. In a context such sensible as frauds tracking, the domain experts have to carefully check and label new positives before the training set be updated.

In order to test the initialization performance, we use a small training dataset representing 1.3% of the distinct root values in the dataset. The test dataset is reduced to 60,000 values containing 0.0001% of positive values, in order to simulate a real-world situation. The arbitrarily chosen dimension of $IG(N, A)$ is 7. After tuning, DRIN reaches 100% recall with 40% precision.

We observe falloffs of about 20% in recall with larger test datasets of about six million values. We observe a significant effect of parameters tuning with large datasets which makes difficult to ensure that the best setting is found. Increasing the training set to 70% of the distinct root values, we observe no significant changes in recall but the precision raises up to 55%, with an incomplete tuning due to time lack. These statistical observations show a remarkable resistance of DRIN to initialization with a tiny training dataset. However, one must be careful with the relevance of certain statistical measures in line with certain application purposes, such as for weak signals, fraud detection related to facsimiles cases or the detection of unexpected and/or unknown events. *F-score* (precision and recall) requiring exact data, it turns irrelevant with corrupted training sets. By the way, it compares retrieved patterns or properties with experts-defined labeled objects and so, does not take into account undefined objects. Therefore it is not convenient for measuring the detection of unknown patterns or properties.

As an example, the most important observation to retain in accordance with the main goal of DRIN, comes from a double blind test realized during an experimentation phase. A few positives, presenting a rare particularity to the domain experts, were incidentally deleted from the training dataset/hidden to the system while their related data, initially not quantified with any weights as all initial data, were left intact in the test dataset.

As DRIN is not a discrete classifier but a continuous one, it ranks every elements with a risk probability in $OG(N, A)$. With several couples $[usd; OG(N, A)]$, DRIN found the same Top 3 ranked identifiers. This particularity having called the attention, domain experts were solicited and investigations revealed the anomaly.

Despite of endogenous data tied to the missing positives had not been highly ranked as it is for known positives, these particular missing positives have been detected as a principal risk without appearing in the training dataset. Statistical hypothesis testing appears not reliable to proof and explain such a case. May be due to the multidisciplinary foundations of DRIN, the missing positives might have been estimated as a systemic anomaly (*i.e.*, a physical aberration/irregularity)[6] in the dataset $\{DS\}$, triggering a ranking alert.

6 Conclusions and Perspectives

Deep Resonance Interference Network (DRIN) results from a multidisciplinary study. It defines deep probabilistic networks aiming at the simulation of time reversal in a discrete and finite space [22] to provide a solution for frauds detection related to the facsimile problem—introduced beforehand—in Big Data.

The proposed model is not affected from the rapid changes of organized attacks and patterns (rules-free system) and let the experts focus on visual frauds cases and/or risks instead of chasing unknown rules in perpetual change. It concerns Network Science, a multidisciplinary branch of Data Science defined since 2005 by the United States National Research Council [35].

DRIN defines a probabilistic feed-forward and back-propagation process inspired from wave-based time reversal within deep heterogeneous networks. It generates artificial probability values within a coherent and multidimensional data space without global attenuation or amplification. As a result, it makes appear deep local amplification/absorption of conjugated probabilities as a signal focused on subtle fraud cases invisible to the domains experts, which are by nature the most efficient ones. The propagation phase of DRIN simulates a discrete wave field and resonances or interferences from a training signal in a multidimensional and coherent space. The back-propagation phase simulates a reversal as an echo of the training signal within the wave field, for a better separation of resonances and interferences in final representations resulting from the model.

Successive experiments were realized with several databases from about 1 to 100 GB including data sets up to 1000 variables mapped onto one billion rows, since the first stages. Concomitant spaces were explored as distinct communities, thanks to social networks analysis and mining [26–28, 30].

Statistical observations show a remarkable resistance to initialization from a tiny training dataset—*e.g.*, 1.3% of distinct test values reach 100% recall. Mean precision were observed with large test datasets (55% for six million rows), may be due to a lack of optimized parametrizing. However, interesting but not measurable outcomes were observed in an double-blind experiment.

Particular positives have been detected as a major risk, without appearing in the training dataset. There is still no clear demonstration of such a finding. It might be due to the multidisciplinary foundations of DRIN from which missing positives should be estimated as a systemic anomaly—*i.e.*, a physical aberration/irregularity.

[6] An aberration acting "as a black hole" in $IG(N, A)$.

DRIN works on small hardware clusters with small/big data sets. Experimentation with a 176 cores-1 TB hardware cluster and other data sets, up to 30 TB, is under preparation. Future works should be focused on optimizing multiple parameters online, for automatic tuning related to user-defined choices, and on the perspective of triggering real-time alerts to support the domain experts.

Finally, neuronal networks models such as ANN, GTN, CNN, RNN are all deep networks structures stemming from optimization and separation functions. Their biomimetic inspiration, historically anchored in the biological observation of visual cortex, shows through their resulting applications that gravitate around the recognition of visual patterns in images, text or streams and around perception—cf. Perceptron.

A first alternative in deep learning recently appeared with adversarial networks and their applications in imagery and artistic creation. Deep Resonance Interference Network proposes another alternative in the field of deep networks and machine learning, modifying the principle *optimizer-separator*, inherent to the current models in deep learning, into a novel principle *correlator-unifier*, toward an applicative field closer to reasoning and cognitive processes than to perception. Instead of recognizing known patterns from multiple features learned from numerous data, it enables to detect unknown patterns from a unique feature learned from sparse data. As a result, it might open a new track in AI for discovering linked features in big data such as exposed from the social and semantic web and so, foresee future applications in numerous fields for developing AI as *societal learning* for mankind.

References

1. H. Putnam, *Mind, Language and Reality.* (Cambridge University Press, Cambridge, 1975)
2. E. Balystok, G. Poarch, Language experience changes language and cognitive ability. Z. Erziehwiss. **17**(3), 433–446 (2014)
3. M. Corballis, From mouth to hand: Gesture, speech, and the evolution of right-handedness. Behav. Brain Sci. **2**(26), 199–208 (2003)
4. A.M. Turing, Computing machinery and intelligence. Mind **59**(236), 433–460 (1950)
5. J. Pitrat, *Métaconnaissance: Futur de l'intelligence artificielle* (Hermès, Hoboken, 1990)
6. J. Pitrat, *Artificial Beings: The Conscience of a Conscious Machine*, ISTE (Wiley, Hoboken, 2010)
7. Y. Lecun, L. Bottou, Y. Bengio, P. Haffner, Gradient-based learning applied to document recognition, in *Proceedings of the IEEE* (1998), pp. 2278–2324
8. C. Dimitrakakis, C.A. Rothkopf, Bayesian multitask inverse reinforcement learning, in *Recent Advances in Reinforcement Learning. EWRL 2011*, vol. 7188, ed. by S. Sanner, M. Hutter. Lecture Notes in Computer Science (Springer, Berlin, 2012)
9. A. Krizhevsky, I. Sutskever, G. Hinton, ImageNet classification with deep convolutional neural networks, in *Advances in Neural Information Processing Systems 25: 26th Annual Conference on Neural Information Processing Systems 2012. Lake Tahoe, Nevada* (2012), pp. 1106–1114
10. A. Galván, Neural plasticity of development and learning. Hum. Brain Mapp. **31**(6), 879–890 (2010)
11. L. Chen, L. Qi, Social opinion mining for supporting buyer's complex decision making: exploratory user study and algorithm comparison. Soc. Netw. Anal. Min. J. **1**(4), 301–320 (2011)

12. Z. Ghahramani, Probabilistic machine learning and artificial intelligence. Nature **521**, 452–459 (2015)
13. R. Brause, T. Langsdorf, M. Hepp, Neural data mining for credit card fraud detection, in *Proceedings of the 11th IEEE International Conference on Tools with Artificial Intelligence, ICTAI '99* (IEEE Computer Society, Washington, 1999), p. 103
14. C. Thovex, Deep probabilistic learning in hidden social networks and facsimile detection, in *The 2018 IEEE/ACM International Conference on Advances in Social Networks Analysis and Mining, ASONAM2018, Barcelona* (IEEE Computer Society, Washington, 2018), pp. 731–735
15. D. Hubel, T. Wiesel, Receptive fields, binocular interaction and functional architecture in the cat's visual cortex. J. Physiol. **1**(160), 106–154 (1962)
16. Q. Wang, Artificial neural network and hidden space SVM for fault detection in power system, in *ISNN 2009: Proceedings of the 6th International Symposium on Neural Networks* (Springer, Berlin, 2009)
17. I. Goodfellow, J. Pouget-Abadie, M. Mirza, B. Xu, D. Warde-Farley, S. Ozair, A. Courville, Y. Bengio, Generative adversarial nets. in *Advances in Neural Information Processing Systems*, vol. 27, ed. by Z. Ghahramani, M. Welling, C. Cortes, N.D. Lawrence, K.Q. Weinberger (Curran Associates, Inc, Red Hook, 2014), pp. 2672–2680
18. J. Von Neumann, *Theory of Games and Economic Behavior* (Princeton University Press, Princeton, 1944)
19. J. Starck, F. Murtagh, J. Fadili, *Sparse Image and Signal Processing: Wavelets and Related Geometric Multiscale Analysis* (Cambridge University Press, Cambridge, 2015)
20. A. Tarantola, B. Valette, Generalized nonlinear inverse problems solved using the least squares criterion. Rev. Geophys. **20**(2), 219–232 (1982)
21. K. Mosegaard, A. Tarantola, Probabilistic approach to inverse problems, in *in the International Handbook of Earthquake & Engineering Seismology (Part A)* (Academic, Cambridge, 2002), pp. 237–265
22. M. Fink, Time reversal of ultrasonic fields. i. basic principles. IEEE Trans. Ultrason. Ferroelectr. Freq. Control **39**(5), 555–566 (1992)
23. S. Brin, L. Page, The anatomy of a large-scale hypertextual web search engine, in *Proceedings of the seventh International Conference on the World Wide Web (WWW1998)* (1998), pp. 107–117
24. A. Yessad, C. Faron, R. Dieng, T. Laskri, Ontology-driven adaptive course generation for web-based education, in *ED-MEDIA 2008 (World Conference on Educational Multimedia, Hypermedia and Telecommunications, Vienna, June 30–July 4, 2008)* (2008)
25. J.L. Moreno, *Who Shall Survive: A New Approach to the Problem of Human Interrelations* (Nervous and Mental Disease Publishing Co., New York, 1934)
26. L. Freeman, W. Bloomberg, S. Koff, M. Sunshine, T. Fararo, *Local Community Leadership* (Syracuse, New York, 1960)
27. L. Freeman, D. White, A. Romney, *Research Methods in Social Network Analysis.* (George Mason University Press, Fairfax, 1989)
28. M. Newman, Detecting community structure in networks. Eur. Phys. J. B., **38**(2), 321–330 (2004)
29. A. Koppal, The Ising Model and Percolation on Graphs (2008). http://www1.cs.columbia.edu/~coms6998/Notes/lecture22.pdf
30. C. Thovex, B. Le Grand, O. Cervantes, A.J. Sánchez, F. Trichet, *Encyclopedia of Social Network Analysis and Mining*, Semantic Social Networks Analysis (Springer, New York, 2017), pp. 1–12
31. P. Wilinski, B. Solaiman, A. Hillion, W. Czarnecki, Toward the border between neural and Markovian paradigms. IEEE Trans. Syst. Man Cybern. B **28**(2), 146–159 (1998)
32. L.E. Mukhanov, Using Bayesian belief networks for credit card fraud detection, in *Proceedings of the 26th IASTED International Conference on Artificial Intelligence and Applications, Anaheim* (ACTA Press, Calgary, 2008), pp. 221–225

33. M. Pearson, P. West, Drifting smoke rings: social network analysis and Markov processes in a longitudinal study of friendship groups and risk taking. Connections Bull. Int. Netw. Soc. Netw. Anal. **25**(2), 59–76 (2003)
34. L. Saul, M. Jordan, Mixed memory Markov models: Decomposing complex stochastic processes as mixtures of simpler ones. Mach. Learn. **37**, 75–87 (1999)
35. N.R. Council, *Network Science* (The National Academies Press, Washington, 2005)
36. J.G. Gottling, Node and mesh analysis by inspection. IEEE Trans. Educ. **38**(4), 312–316 (1995)
37. J. Rosenschein, L. Pack Kaelbling, The synthesis of digital machines with provable epistemic properties, in *Proceedings of the Conference on Theoretical Aspects of Reasoning About Knowledge*, ed. by J. Halpern (Morgan Kaufmann, Burlington, 1986), pp. 83–98
38. C. Thovex, F. Trichet, An epistemic equivalence for predictive social networks analysis, in *Web Information Systems Engineering – WISE 2011 and 2012 Workshops, Sydney, Australia and Paphos, Cyprus*, vol. 7652 Lecture Notes in Computer Sciences (LNCS) (Springer, Berlin, 2013), pp. 201–214.
39. L. Bottou, From machine learning to machine reasoning. Mach. Learn. **94**(2), 133–149 (2014)
40. P. Erdõs, A. Rényi, On random graphs. Publ. Math. **6**, 290–297 (1959)

User's Research Interests Based Paper Recommendation System: A Deep Learning Approach

Betül Bulut, Esra Gündoğan, Buket Kaya, Reda Alhajj, and Mehmet Kaya

1 Introduction

The information in the world of science increases constantly. Scientists constantly make new scientific studies. Academic studies are converted into publications such as books, journals and conference proceedings. This rapid development in science has led to a rapid increase in the number of publications in the digital media. Researchers follow the studies about their study fields. Numerous databases are available to help researchers access information. The databases like Google Scholar, Science Direct and IEEE Digital Library are the most preferred databases especially in engineering. Increasing the number of digital libraries over the years and increasing the number of publications covered by the existing databases is a positive development for scientists. It is a positive situation for the researcher to have so many sources of information. However, it is difficult for the researcher to choose publications that may be useful to him in an excessive amount of publications in different sources. Filtering is essential for the researcher to find the most suitable publication from a database full of publications.

B. Bulut · E. Gündoğan · M. Kaya (✉)
Department of Computer Engineering, Fırat University, Elazığ, Turkey
e-mail: kaya@firat.edu.tr

B. Kaya
Department of Electronics and Automation, Fırat University, Elazığ, Turkey
e-mail: bkaya@firat.edu.tr

R. Alhajj
Department of Computer Science, University of Calgary, Calgary, AB, Canada

Department of Computer Engineering, Istanbul Medipol University, Istanbul, Turkey
e-mail: alhajj@ucalgary.ca

M. Kaya et al. (eds.), *Putting Social Media and Networking Data in Practice for Education, Planning, Prediction and Recommendation*, Lecture Notes in Social Networks, https://doi.org/10.1007/978-3-030-33698-1_7

There are article recommendation systems for researchers to facilitate publication. The most important aim of the article recommendation is to identify the articles that are closest to the researcher. Another aim is to minimize the time spent by the researcher to find the publication in which he or she is interested.

Most of article recommendation studies compare the content of the candidate articles. In the database, the articles that are most similar to each other are recommended to each other. These studies do not consider the researcher. The same articles are recommended to all researchers without paying attention to the researcher's previous studies, the study field. However, an article may not attract the interest of all researchers studying in that field.

In classical article recommendation systems, the researcher's publications or the study field is of no importance. The recommendation is made by accepting each researcher equally. In the proposed method, article recommendation is made by considering the features of the researcher. The difference of our study from other studies is that each user recommends the publication that is closest to the field of study taking into account the features of each researcher.

Our study aims to create a user-specific article recommendation system by considering the features such as the publications in the researcher profile and study field.

In our study, a recommendation system was created by using IEEE database data. It was performed with multiple similarity algorithms. The comparison of the user profile with candidate article to be recommended was first performed with the cosine similarity algorithm. Then, it was repeated by Doc2vec method which is one of the deep learning methods. It was tested with the voluntary participation of researchers with different degrees who work in different fields. It was seen that our study was more successful than other studies.

The rest of the paper is organized as follows. Section 2 offers an overview of article recommendation approaches. Section 3 describes the basic architecture and functioning of the system. Section 4 discusses results obtained from the recommendation system. Section 5 explains the conclusions.

2 Literature Review

Articles are prepared and published as text. Therefore, we can say that all studies conducted for the purpose of article recommendation are text processing or text mining. The common feature of all studies in the literature is that they do text processing. We used cosine similarity, TF-IDF and Doc2vec methods in our study. Doc2vec is a very popular machine learning method using recursive neural networks. It pays attention to the semantic meanings of words in text processing and considers synonymous words in text similarity studies. Deep learning methods provide more advanced technology and better results than classical machine learning methods.

There are many studies on article recommendation. Some of these studies are content-based, some are labeling-based and some are hybrid.

Although there have been many studies on labeling, the number of studies dealing with academic articles is relatively low. One of the outstanding studies on labeling and academic article recommendation method is the one proposed by Choochaiwattana [1] in 2010, and the other is Bahulikar's [2] study in 2017. Contrary to the known methods Ravi et al. [3] have proposed a model has that process text using recurrent neural networks and recommends academic articles accordingly. In addition, ontology-based academic article recommendation studies have also been conducted [4, 5].

There are also hybrid studies within academic article recommendation studies. While Lee et al. [6] have proposed a new model combining 2 different approaches, content-based and graph-based; Bancu et al. [7] have proposed a study that combines content-based and collaborative filtering methods. Apart from these, there are also approaches that combine two sets of data and use it as an academic article recommendation method, as is done by Zhao et al. [8].

West et al. [9] have indicated the references with nodes. They sorted the references with the Eigenfactor algorithm and clustered the knots using MapEquation. They found the most important node in the different steps of the hierarchy of these clusters. They have used an algorithm similar to PageRank to find this node. They have recommended the article of the most important node they found.

Xia et al. [10] have conducted a study that recommended articles with same authors for articles containing more than one author. Zhou et al. [11] have tried to list the confidence index of each academic article by suggesting a paper rank algorithm similar to the page rank algorithm, different from all studies and constructed the recommendation system on the data obtained from this. Drushku et al. [12] have proposed a collaborative method based on user interactions in their study. They set similarity criteria using a set of clustering techniques by looking at the user modelling of search engines. This study is similar to our study because similarity algorithms are successful and pay attention to the user profile.

Finding similarity of different articles with the keyword can be called a label-based similarity computation algorithm. Although researchers have suggested many methods in past studies, the most simple and easily integrated TF-IDF method was used for this method. There are many other studies [13–17] that use this method for the same purpose with us.

There are also similar recommendation systems [18–20] using bipartite graphs. Ohta et al. have proposed an academic article suggestion model using bipartite graphs.

Ashraf et al. [21] have taken the data from users on social media and recommended them according to the interest of the users. This study is similar to our study because it offers recommendation based on the user's profile. However, while our study recommends articles, this study recommends news. Our study is more complicated. It is more important because of its contribution to the advancement of science.

Watanabe et al. also have collected the information like the number of times an academic article was read except for its metadata and added them to the system of recommendation [22]. Similar to our study, the academic article recommendation model of Cui et al. [23] that also deals with the relationship among the users is one of few studies on this topic.

Xue et al. and Chen et al. have suggested one of the most related studies to our study because they use both online databases and analyze the user's published studies to determine the user's interest [24, 25]. The most original feature of the study of Xue et al. is that it implements the algorithm in a real online academic database. Chen and his colleagues have suggested that researchers should work with the vision that they can specialize in more than one field because of the fact that there is not such a limitation for them to specialize in just one field.

Bai et al. [26] have made a comprehensive review of the article recommendation systems. They have compared content-based, graphic-based methods, filtering methods and hybrid methods. Du et al. [27] have used the LDA (Latent Dirichlet Allocation) model to develop an article recommendation system based on their collaboration with co-authors of the articles. The study combining the Hadoop method with the LDA method is more successful than the collaborative filtering algorithm. However, this study does not pay attention to the other features of the articles. Therefore, our study is more successful than this study. Li et al. [28] have proposed an approach to e-commerce applications by including the user's social knowledge. Also they have created an effective confidence measurement model. Hebatallah and Hassan [29] have proposed a personalized research recommendation system based on user's explicit and implicit feedback. They have used recurrent neural networks in their studies. The difference between this study and our study, this study is based on the researcher's feedback, while our study looks at the researcher's publications. Karvelis et al. [30] have proposed the topic by using the Doc2vec method and the bag of words method. They have aimed to obtain a set of correct topic terms for library books. Similar to our study, the study is to use the Doc2vec method for recommendation. Our study is more comprehensive than this study as it recommends according to the user's profile. Rabindra Nath Nandi et al. [31] have applied Doc2vec method in their study. This study compares the results of Bag of words, LDA and LSA methods. Similar to our study, Doc2vec is used. The text is processed and recommended.

There are many studies on deep-learning text processing. Lee et al. [32] have made deep learning predictive text with deep learning. In the study, a film scenario has been created and the events that would be in the next scene with emotion analysis has predicted as text. Target and predicted emotion changes have been tested with cosine similarity. This study is similar to our study in terms of using Deep-Learning. The study by Qu et al. [33] aims to recommend on deep social media platforms with deep learning methods. The study makes friends recommendation using deep learning methods. Deep Graph-Based Neural Network (DGBNN) framework has been proposed.

3 Context

In this study, the most frequently used digital databases were used by the researchers in Computer Sciences, Electrical and Electronics Engineering and Mechanical Engineering to create the data set. Academic databases used as data set are:

1. Association for Computing Machinery (ACM) Digital Library
2. IEEE Xplore Digital Library
3. The DBLP Computer Science Bibliography

ACM Digital Library was used for the metadata such as title, author name, keyword, article date and abstract. IEEE Xplore Digital Library was used for the metadata such as title, author name, keyword, article year and abstract. The DBLP Computer Science Bibliography digital library was used for the metadata such as article year, author name, article title. IEEE Xplore Digital Library and ACM Digital Library data has never been stored in our system. When we make a user-based recommendation for the article, only the selected article, the user's publications, and the public information of the articles containing the same keyword with the selected article were taken over HTTP. This information was not kept after the recommendation. Article year, article title and abstract metadata were placed in separate sections in our temporary data repository according to the articles they belonged. As author name is the most important point in our article recommendation system, it is shown both in the metadata table and a separate table.

4 Methodology

According to user profile information, the article recommendation system consists of four basic structures. The data from the datasets are transmitted to the dataset interpreter. The dataset interpreter separates the user information data from the ACM Digital Library, the IEEE Xplore Digital Library, and The DBLP Computer Science Bibliography datasets. It parses the metadata of each article. It also parses and rearranges the metadata of the articles that is recommended as a result of the searched keyword. The parsed data are kept in Datasep Interpreter. Figure 1 shows the structure of the system architecture.

In the system that recommends article according to the user profile information, we can basically divide the data from the datasets into two parts:

1. A set of metadata that is in the researcher's profile including year, title, abstract, keywords of each of the articles that they have so far.
2. A set of metadata including year, title, abstract, keywords of the articles related to the researcher's field.

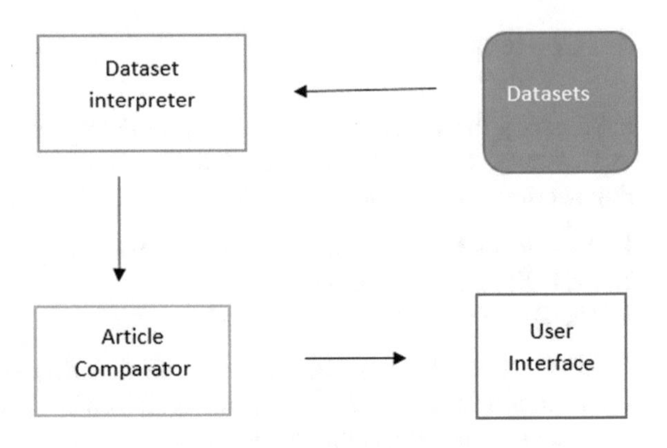

Fig. 1 System architecture

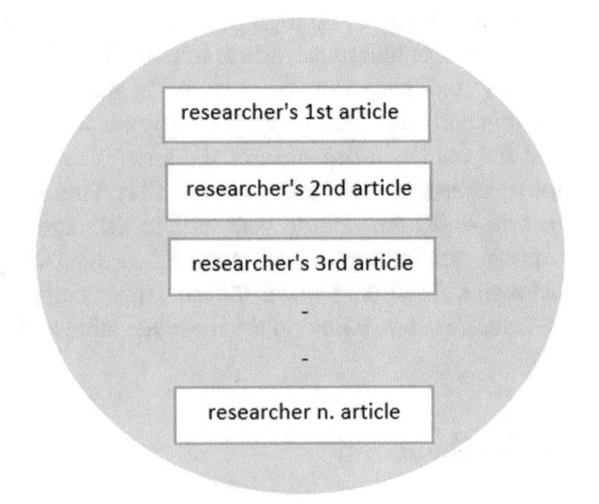

Fig. 2 Creating a profile from author's articles

In Fig. 2, the metadata of each article previously written by the researcher are combined. In this way the user profile is created.

$$info_n = title_n + year_n + authors_n + abstract_n + keywords_n \quad (1)$$

Equation (1) represents the sum of the metadata information of article number n. The formula that contains the metadata information of all the author's articles is as follows:

$$profile\_info = \sum_{k=0}^{n} info_n \quad (2)$$

Equation (2) is a string that is the sum of metadata information for all articles of the author. In other words, it is the profile information of the user. The profile_info value is the profile of the author whose article is to be recommended. $info_n$ value is the metadata information calculated for each article of the author expressed in Eq. (1).

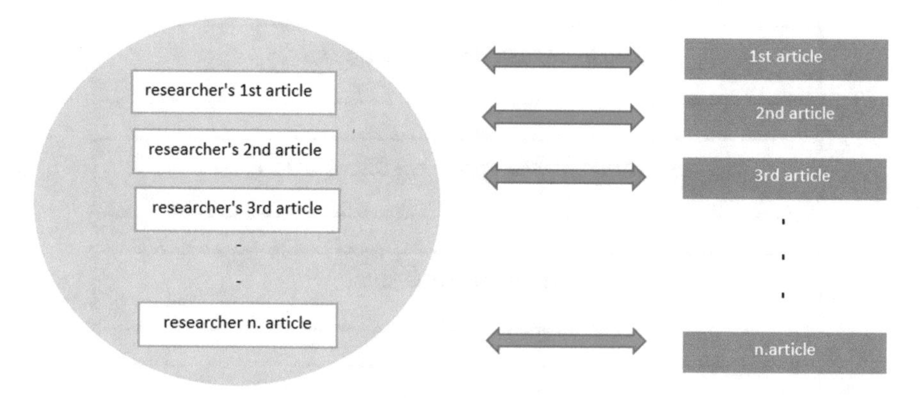

Fig. 3 Author profile and article comparison

The user's profile is created by combining the information of the n articles written by the researcher. The created profile and each of the articles are compared one by one. Figure 3 represents the comparison processes.

The processed data of the author's information and the results from the keywords are compared to the article. It compares the articles using comparative cosine similarity and TF-IDF methods. The article comparator compares the author information (*profile_info*) and each of the articles ($info_n$). The comparison result score of each article is determined. According to the comparison results from the comparator, the most related ten articles are presented to the user through User Interface.

We realized our study with more than one method. With the cosine similarity and TF_IDF method, we recommended article to the researcher and got the results. The second method of our study is the Deep Learning method which is the use of advanced technology, multi-level deep neural networks. It is very popular in high technology applications. As with machine learning methods, deep-learning method also learns system with training data. According to these data, it is obtained from the test data the results.

Gensim 3.4 and python 3 are the versions of the technologies used in our project. The Doc2vec method was used to find the most compatible article with the user's profile. For this purpose, the user's profile was created first. The user's profile consists of a combination of metadata such as the title, keyword and abstract of the articles of the user.

The Word2vec method breaks the given text into words and each word changes to a vector. In Word2vec, each word is a vector and the angle between the vectors of the words is looked at. These two words are similar to each other if the angle of the two vectors is close to each other.

The Doc2vec method is basically created based on the Word2vec method. However, in Word2vec, each word is expressed in a vector, while in Doc2vec, each text or document is a vector. The similarity between these two texts is determined

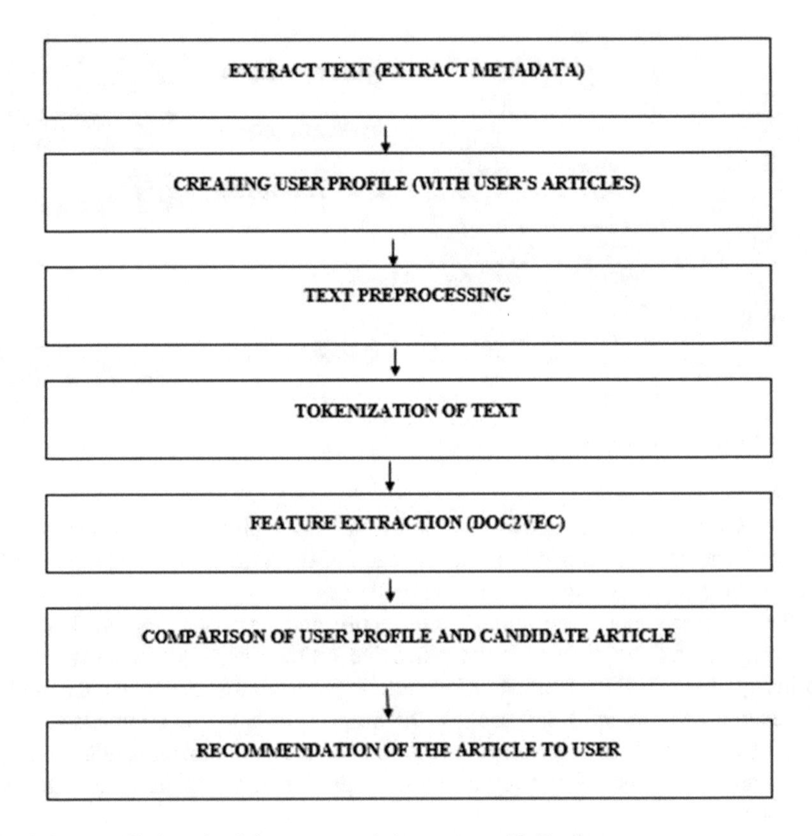

Fig. 4 User profile-based article recommendation system with Doc2vec

by looking at the degree of angle between the vectors of two texts. Figure 4 is the architecture of the article recommendation system developed with Doc2vec. The first step of the Doc2vec method is to split the text into tokens. Natural Language Toolkit platform was used for this in the study. The text containing the information obtained from the user's profile was divided into token by the NLTK tokenizer. The Doc2vec model was created with the data allocated to the tokens.

Vector_size refers to the dimensionality of the feature vectors in the Doc2vec method. In our study, we studied with more than one vector_size value. The most successful value was found to be vec_size = 10. The variable min_alpha is a float variable. It was seen that the learning rate decreased linearly to min_alfa during the study. Alpha is the first learning rate. The alpha value was selected as 0.025. The value of the epochs is the number of occurrences of the data on the corpus as the model is created. The epoch value was given as 100. The values selected for the variables are the values from which the most successful results are obtained.

All the words letters of the text we obtained from the metadata of the articles in the user profile reserved for the tokens were converted to lowercase letters. The

Doc2vec model was created with the obtained data and was recorded. The Doc2vec model obtained here can be defined as the vector of the user profile.

IEEE database was searched with the keyword taken from the researcher and the information such as title, author, keyword, and abstract of the articles was obtained. This information was indexed as elements of an array. Metadata was divided into token as it was when creating a user profile model. The data allocated to the tokens were converted to lowercase letters. The metadata of the article was converted to vector by Doc2vec method. In our study, each of these articles in the search results was expressed with a different vector. So, each element of the array was a separate vector.

The user profile vector and the vector of the article vectors are compared with the Doc2vec method. The vectors close to the user profile vector were articles that were recommended to the user. The comparison was performed with the most similar method of the Gensim library. This method takes the value of top-n as a parameter and the vector that is most similar to the number given at the value of the top-n is taken. In our study, we gave the value of top-n 10. We recommended 10 articles that were most similar to the user's profile.

5 Results

There are only a few studies that make article recommendations by paying attention to the information of the articles in the user's profile. In this study, what we tried is to determine the success criterion of the proposed method by using a few of the previously proposed methods.

In the first method, random email addresses and names of researchers were determined as experimental groups. The system sent the links and titles of ten articles that may interest the researchers based on their articles. The number of emails sent is ten, because it was anticipated that the researcher may not click on all links when a lot of articles are sent.

Table 1 shows the number of mails taken by the researcher, the number of mails read by the researcher, and the number of the researchers' own articles. The researchers read 47% of the mails and read 29% of the articles.

Another method is to conduct a survey with a research group. Name and surname of the researchers who will participate in the survey were taken. Recommended articles created from our system were taken from the system. Our system recommended 20 articles specific to each researcher. Each researcher who participated in the survey was presented the articles compiled for themselves in the form of a survey. In this two-choice survey, the researchers were asked to mark yes if the article interest them and mark no if not.

Table 2 shows the number of researchers participating in the experiment, the number of articles sent to the researchers and the answers of the researchers to this survey. The researchers have marked the articles that interest and do not interest them in the survey. According to the results, 426 of the articles submitted to the

Table 1 Experimental results

Number of emails sent to the researcher	Number of emails read by the researcher	Number of publications clicked by the researcher	Number of researcher's own publications
10	7	6	52
10	3	1	18
10	8	5	26
10	5	4	15
10	4	2	20
10	3	2	15
10	6	3	31
10	1	1	19
10	7	4	37
10	3	1	17

Table 2 Numbers of researchers and articles

Number of researchers participating in the survey	Number of articles presented in the survey	Number of articles marked as "Yes" by researchers in the survey	Number of articles marked as "No" by researchers in the survey
30	600	426	174

researcher were marked as yes, 174 as no. 71% of the articles attracted the interest of the researchers, 29% did not.

The Word2vec method represents a word with a vector. The Doc2vec method works on the same principle as the Word2vec method. Both methods are based on the semantic features of the words. Compared to the Word2vec method, the Doc2vec method can represent a whole sentence with a vector. In our study with the Doc2vec method, we showed each paragraph with a vector.

Doc2vec has two different models [31]. The first of these models is distributed memory model. This model is shown in Fig. 5.

In the Distributed Memory Model, D is the paragraph vector. W is the vector of words. Paragraph vector and word vectors are averaged. The fourth word is predicted with this average. The second model is the Distributed Bag of Words model. This model is shown in Fig. 6.

Paragraph1: "Wheather defines the state of the atmosphere. The weather is cold, rainy, dry, calm or windy. Air is a word that expresses all events taking place in the atmosphere layer."

Paragraph2: "All the natural phenomena occurring in the sky are air. All events such as rain, snow, cold, hot tornado, storm lightning define the weather."

Paragraph3: "Agriculture is the name given to the team to get the harvest of agricultural land. One of the most important branches of agriculture is organic agriculture."

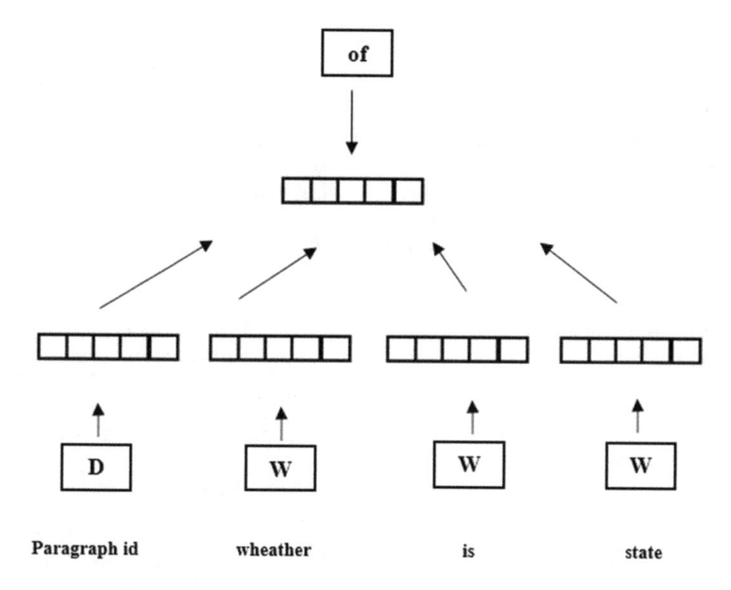

Fig. 5 The framework for distributed memory model

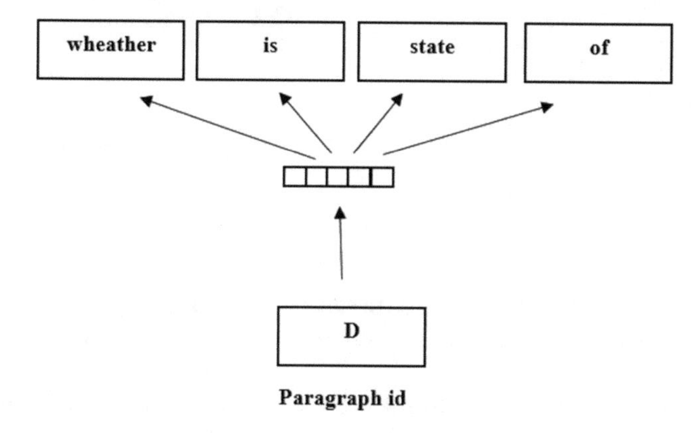

Fig. 6 The framework for distributed bag of words model

The above example is an example where the first and second paragraph should be closer to each other than to the third paragraph. In the Doc2vec model we created, these 3 paragraphs were compared. It was found that the first two paragraphs were similar to each other. In this model, the vectors of the words were combined to obtain the vector of the paragraph. We used Doc2vec's Distributed Bag of Words method in our study. The reason for this selection is that the Distributed Memory Model is useful in predicting missing words. However, there must be a vector of each paragraph to recommend article. The closest article is recommended to the researcher by comparing these vectors.

Table 3 Similarity score of two sample text in Doc2vec

Sample text I	Sample text II	Similarity score
Structured Query Language is a language that is designed to manage data stored in relational database management systems and is specific to an area being programmed	SQL is definitely not a programming language. It is used for database operations as a sub language. SQL is used to process data stored in the database. Enables data to be modelled within the intended task definition	0.363

Table 4 Similarity scores of articles with Doc2vec

User	The similarity score of 5 recommendation article per user
User1	0.727
User2	0.245
User3	0.123
User4	0.316
User5	0.211
User6	0.339
User7	0.218
User8	0.174
User9	0.120

Table 5 Success rates of Doc2vec method

Accuracy	83.4%
Precision	78.6%
Recall	47.2%
Specificity	96.7%
F1-measure	55.3%

In this study, the text obtained from the user profile consisting of 1805 words was used as the training data. The results of vector similarity for a sample text are shown in Table 3.

In this study, we selected 9 different authors and recommended 5 articles for each author. The average of similarity scores with the user profile of the 5 recommended articles is given in the Table 4.

The success rates of the Doc2vec method tested according to the survey method are shown in Table 5. According to Table 5, the accuracy rate of the system is 83.4%.

6 Conclusions

A In this study, ACM Digital Library, IEEE Xplore Digital Library, and The DBLP Computer Science Bibliography are used. The system has been tested by sending mail to the researchers. E-mails are sent in two ways. First, the researchers were sent e-mails and asked to click on these e-mails. Forty-seven percent of the researchers clicked on emails. In the second method, questionnaires were sent to the researchers via e-mail.

According to the survey technique, 71% of the emails attracted the attention of the researcher. According to these results, the survey technique is more successful than the technique based on clicking on the e-mails. Doc2vec method is a successful method for our article recommendation system with 83.4% accuracy. Unlike classical article recommendation systems, the study evaluates publications by the researcher and recommends publications that may be of interest to the researcher. The results of the system that make it are promising.

References

1. W. Choochaiwattana, Usage of tagging for research paper recommendation, in *ICACTE 2010 - 2010 3rd International Conference on Advanced Computer Theory and Engineering, Proceedings*, vol. 2 (IEEE, 2010)
2. S. Bahulikar, Analyzing recommender systems and applying a location based approach using tagging, in *2017 2nd International Conference for Convergence in Technology (I2CT)* (IEEE, Piscataway, 2017)
3. K.M. Ravi, J. Mori, I. Sakata, Cross-domain academic paper recommendation by semantic linkage approach using text analysis and recurrent neural networks, in *2017 Portland International Conference on Management of Engineering and Technology (PICMET)* (IEEE, 2017)
4. S.S. Weng, H.L. Chang, Using ontology network analysis for research document recommendation. Expert Syst. Appl. **34**(3), 1857–1869 (2008)
5. K.V. Neethukrishnan, K.P. Swaraj, Ontology based research paper recommendation using personal ontology similarity method, in *2017 Second International Conference on Electrical, Computer and Communication Technologies (ICECCT)* (IEEE, 2017)
6. Y.C. Lee et al., Recommendation of research papers in DBpia: a Hybrid approach exploiting content and collaborative data, in *2016 IEEE International Conference on Systems, Man, and Cybernetics, SMC 2016 - Conference Proceedings* (IEEE, 2017), pp. 2966–2971
7. C. Bancu, M. Dagadita, M. Dascalu, C. Dobre, S. Trausan-Matu, A.M. Florea, ARSYS—article recommender system, in *Proceedings - 14th International Symposium on Symbolic and Numeric Algorithms for Scientific Computing, SYNASC 2012* (IEEE Computer Society, 2012), pp. 349–355
8. W. Zhao, R. Wu, H. Liu, Paper recommendation based on the knowledge gap between a researcher's background knowledge and research target. Inf. Process. Manag. **52**(5), 976–988 (2016)
9. J.D. West, I. Wesley-Smith, C.T. Bergstrom, A recommendation system based on hierarchical clustering of an article-level citation network. IEEE Trans. Big Data **2**(2), 113–123 (2016)
10. F. Xia, H. Liu, I. Lee, L. Cao, Scientific article recommendation: exploiting common author relations and historical preferences. IEEE Trans. Big Data **2**(2), 101–112 (2016)
11. Q. Zhou, X. Chen, C. Chen, Authoritative scholarly paper recommendation based on paper communities, in *Proceedings - 17th IEEE International Conference on Computational Science and Engineering, CSE 2014, Jointly with 13th IEEE International Conference on Ubiquitous Computing and Communications, IUCC 2014, 13th International Symposium on Pervasive Systems* (IEEE, 2015), pp. 1536–1540
12. K. Drushku, J. Aligon, N. Labroche, P. Marcel, V. Peralta, Interest-based recommendations for business intelligence users. Inf. Syst. **86**, 79 (2019)
13. W. Zhang, T. Yoshida, X. Tang, A comparative study of TF$_*$IDF, LSI and multi-words for text classification. Expert Syst. Appl. **38**(3), 2758–2765 (2011)

14. J. Ramos, Using TF-IDF to determine word relevance in document queries, in *Proceedings of the First Instructional Conference on Machine Learning* (2003), pp. 1–4
15. T. Kenter, M. de Rijke, Short text similarity with word embeddings categories and subject descriptors, in *Proceedings of the 24th ACM International on Conference on Information and Knowledge Management (CIKM 2015)* (ACM, New York, 2015), pp. 1411–1420
16. S. Albitar, S. Fournier, B. Espinasse, An effective TF/IDF-based text-to-text semantic similarity measure for text classification, in *International Conference on Web Information Systems Engineering*, vol. 8786 (2014), pp. 105–114
17. C.-H. Huang, J. Yin, F. Hou, A text similarity measurement combining word semantic information with TF-IDF method. Jisuanji Xuebao **34**, 856 (2011)
18. A.A. Müngen, M. Kaya, A novel method for event recommendation in meetup, in *2017 IEEE/ACM International Conference on Advances in Social Networks Analysis and Mining (ASONAM)* (ACM, 2017), pp. 959–965
19. E. Gündoğan, B. Kaya, M. Kaya, Prediction of symptom-disease links in online helath forums, in *IEEE/ACM International Conference on Advances in Social Networks and Mining (ASONAM, 2017)* (ACM, 2017), pp. 876–880
20. S. Aslan, M. Kaya, Link prediction methods in bipartite networks, in *2017 International Conference on Computer Science and Engineering (UBMK)* (IEEE, 2017)
21. M. Ashraf, G.A. Tahir, S. Abrar, M. Abdulaali, S. Mushtaq, Personalized news recommendation based on multi-agent framework using social media preferences, in *2018 International Conference on Smart Computing and Electronic Enterprise (ICSCEE)* (IEEE, 2018)
22. S. Watanabe, T. Ito, T. Ozono, T. Shintani, A paper recommendation mechanism for the research support system Papits, in *Proceedings - International Workshop on Data Engineering Issues in E-Commerce, DEEC 2005*, vol. 2005 (IEEE, 2005), pp. 71–80
23. T. Cui, X. Tang, Q. Zeng, User network construction within online paper recommendation systems, in *Proceedings - 2010 IEEE 2nd Symposium on Web Society, SWS 2010* (IEEE, 2010), pp. 361–366
24. H. Xue, J. Guo, Y. Lan, L. Cao, Personalized paper recommendation in online social scholar system, in *ASONAM 2014 - Proceedings of the 2014 IEEE/ACM International Conference on Advances in Social Networks Analysis and Mining* (IEEE, 2014), pp. 612–619
25. J. Chen, Z. Ban, Literature recommendation by researchers' publication analysis, in *2016 IEEE International Conference on Information and Automation (ICIA)* (IEEE, 2016)
26. X. Bai, M. Wang, I. Lee, Z. Yang, X. Kong, F. Xia, Scientific paper recommendation: a survey. IEEE Access **7**, 9324–9339 (2019)
27. G. Du, Y. Liu, J. Yu, Scientific paper recommendation: a survey, in *2018 IEEE Fourth International Conference on Big Data Computing Service and Applications (BigDataService)* (IEEE, 2018)
28. W. Li, X. Zhou, S. Shimizu, M. Xin, J. Jiang, H. Gao, Q. Jin, Personalization recommendation algorithm based on trust correlation degree and matrix factorization. IEEE Access **7**, 45451 (2019)
29. A. Hebatallah, M. Hassan, Personalized research paper recommendation using deep learning, in *UMAP '17 Proceedings of the 25th Conference on User Modeling, Adaptation and Personalization* (ACM, New York, 2017), pp. 327–330
30. P. Karvelis, D. Gavrilis, G. Georgoulas, C. Stylios, Literature recommendation by researchers' publication analysis, in *2016 IEEE International Conference on Information and Automation (ICIA)* (IEEE, 2016)
31. R. N. Nandi et al., Bangla news recommendation using doc2vec. 2018 *International Conference on Bangla Speech and Language Processing* (ICBSLP), pp. 1–5, IEEE. 2018
32. S.-H. Lee, D.-M. Kim, Y.-G. Cheong, Predicting emotion in movie scripts using deep learning, in *2018 IEEE International Conference on Big Data and Smart Computing (BigComp)* (IEEE, 2018)
33. Z. Qu, B. Li, X. Wang, S. Yin, S. Zheng, An efficient recommendation framework on social media platforms based on deep learning, in *2018 IEEE International Conference on Big Data and Smart Computing (BigComp)* (IEEE, 2018)

Characterizing Behavioral Trends in a Community Driven Discussion Platform

Sachin Thukral, Arnab Chatterjee, Hardik Meisheri, Tushar Kataria, Aman Agarwal, Ishan Verma, and Lipika Dey

1 Introduction

The availability of massive amounts of data from electronic footprints of social behavior of humans for a variety of online social networks has triggered a lot of research and its applications. Tools from various disciplines have come together and is currently popular as computational social science [12]. The contemporary approach to social network analysis has improved upon the standard, classical approaches [19] of sociologists, and the present interests span across various disciplines like market intelligence, operations research, survey science, as well as statistical computing. The studies of social network data not only reveals the structure of the connections between its individual components, including their strong and weak ties, and their dynamics, but also the possible reasons as to why those structure and dynamics are prevalent.

A typical social network can be conceived as a multidimensional graph where different elements like users, posts, comments, etc. are the nodes and the links define their interactions. Usual measurable quantities like the lifespan of a post, the average number of posts per unit time etc. reveal aggregate behavior of users across the entire social media platform. User comments across posts render the interactivity flavor,

S. Thukral · A. Chatterjee (✉) · H. Meisheri · I. Verma · L. Dey
TCS Research, New Delhi, India
e-mail: sachi.2@tcs.com; arnab.chatterjee4@tcs.com; hardik.meisheri@tcs.com; ishan.verma@tcs.com; lipika.dey@tcs.com

T. Kataria · A. Agarwal
IIIT Delhi, New Delhi, India
e-mail: tushar15184@iiitd.ac.in; aman15012@iiitd.ac.in

M. Kaya et al. (eds.), *Putting Social Media and Networking Data in Practice for Education, Planning, Prediction and Recommendation*, Lecture Notes in Social Networks, https://doi.org/10.1007/978-3-030-33698-1_8

where user behavior can be segregated according to volume of comments or time span of interaction. The layer of the number of distinct users who are involved, opens a scope to differentiate the behavior of users in terms of impact of the post and its reachability.

Access to huge amounts of data facilitates a rigorous statistical analysis, which can be combined with behavioral studies that can bring out interesting spatial and temporal features, which provides interesting insights. In this article, we study evolution patterns of the posts over time, based on user interactions with the posts and group them into further different categories. We also categorized posts according to user interaction patterns that emerge around them. We present methods to determine the focal points of interactions. Further, we present methods to identify the behavioral trends exhibited by the users to make their posts popular. Additionally, we also discuss methods to analyze the presence of controversial posts and comments, even before getting into the text content.

In *Reddit.com*, users share content in the form of text posts, links and images, which can be voted up/down by other users, where from further discussions may emerge. Posts span over a variety of topics—news, movies, science, music, books, video games, fitness, food, image-sharing, etc. They are organized by subject under user-created *subreddits*, which provide further opportunities for fostering discussion, raising attention and publicity for causes. Reddit is known for its open nature and harbors diverse user community across demographics and subcultures, who generate its content, there is also moderation of posts due to various reasons.

In this article, we gathered insights about where, when and by whom the content is being created in the community as a whole. The study of evolution patterns helps us to understand the characteristics of posts which get large number of responses. Studying these characteristics from an author's perspective gives us an indication on which authors are relatively more reliable in spreading information. Similarly, identifying the focal points of a long discussion may lead to understanding of popular opinions. These markers of behavioral trends can be used as cues in applications like placement of advertisement, summarizing of viral/popular topics from varied perspectives, the half life of information spread, etc.

Social media is being increasingly used for sharing important information across individuals and collaboration, even within enterprises. Understanding human behavior within them and being able to characterize them, as well as to understand the dynamics of interaction within group of users turns out to be an challenging task. For example, in the organization to which most of the authors of this study are affiliated to, more than 400,000 employees engage in at least two organization specific, closed social networks serving different purposes. Analysis of the temporal patterns and the group dynamics presented in our work are important aspects which can not only aid in understanding the different categories of users, but also identify the information needs and push the right content or advertisement for the right group at the right time. The similarity of patterns observed over multiple data sources prove that user behaviors are fairly similar across social platforms in the same domain.

This article is essentially an extension of our recent paper [18] where we presented the primary analysis of behavioral trends observed in Reddit. The rest of the article is organized as follows: We present the earlier related work in Sect. 2. A brief description of the data considered for our study is given in Sect. 3. Analysis of evolution patterns of the posts is reported in Sect. 4. Section 5 reports the interaction dynamics, while Sect. 6 discusses the behavior of authors over the space. Section 7 discusses methods to identify the presence of controversial content in posts and comments. Finally, the entire analysis is summarized along with the inferences in Sect. 8.

2 Related Work

There have been several studies on social media dynamics from various perspectives. Researchers have examined the structure of the comment threads by analyzing the radial tree representation of thread hierarchies [10]. Researchers have also studied the behavioral aspects of users by crowd-sourcing information from experiments performed on the platform. One such study reports how individuals consume information through social news websites, contributing to their ranking systems. A study of the browsing and rating pattern reported that most users hardly read the article that they vote on, and in fact 73% of posts were rated before even viewing the content [8]. While user interactions (likes, votes, clicks, and views) serve as a proxy for the content's quality, popularity, or news-worthiness, predicting user behavior was found to be relatively easy [6]. The voting pattern in the Reddit [14] has been studied to analyze the upvoting of posts from a new page to the front page and behavior of users towards some posts which are getting positive or negative votes. They have studied the posts mentioning Wikileaks and Fox News and to see the impact of negative voting on them, although working on only 1 month of data. One study related to rating effect on posts and comments [7] revealed that random rating manipulations on posts and comments led to significant changes in downstream ratings leading to significantly different final outcomes—positive herding effects for positive treatments on posts, increasing final ratings on the average, but not for positive treatments on comments, while negative herding effects for negative treatments on posts and comments, decreasing the final ratings on average. Another exploratory study [20] on the dynamics of discussion threads found topical hierarchy in discussion threads, and how it is possible to use them to enhance Web search. A study on 'social roles' of users [1] found that the typical "answer person" role is quite prominent, while such individual users are not active beyond one particular subreddit. In another study, authors have used the volume of comments a blog post receives as a proxy of popularity to model the relationship with the text [21]. Authors have used several regression models to predict the volume of comments given in the text. This analysis is quite restricted in terms of the scale of the dataset, limiting to political posts and only three websites which amount to four thousand posts. While the content analysis is most intuitive, it does not provide richer analysis. Text content shared over social media is usually noisy, full of non-standard grammar and spelling, often cryptic and uninformative to the outsider from the community. When one adds the scale of today's social media data, it is computationally non-viable to have content analysis over the whole corpus.

Most of the studies reported till date have performed analysis on a subset of data by restricting themselves to a limited number of posts, comments, top users, subreddits, etc., while we use two separate sets each of which are complete data for 1 year period. To the best of our knowledge, only very few researchers have used complete data for analysis. In Ref. [16], authors have presented evolution analysis over 5 years of subreddits with respect to text, images, and links though they have only considered posts and not analyzed comments. Reference [4] has reported the effect of missing data and its implications over the Reddit corpus taken from 2005 to 2016.

The study of controversies in social media platforms is important in several contexts. The earliest papers that dealt with the issue mainly focused on controversies arising in political conversations, in a typical setting of U.S. Presidential elections, from political news and blogs as well as posts in Twitter. There have been reports of high correlation between negative affect and biased language with controversial issues [13]. In a study of Twitter data, there are reports of highly segregated partisan structure from the retweet graph, with sparse connectivity between the opposite polarities [3]. There exists a body of literature like the above, which make use of the interaction graph structure to uncover and eventually quantify controversies (See also Ref. [5]). However, a text based, sentiment analysis adds to the overall picture, and is also attempted in a few studies [2, 13]. We take a rather simple, rigorous statistical approach, to lay the foundation for future text-based analysis of content. The Reddit corpus is usually devoid of controversial text, as they are removed by moderators, and hence the remaining corpus around the deleted content requires detailed analysis, if one aims at finding indicators of forthcoming controversies.

3 Data Description

3.1 Terminologies

The following are the terminologies that will be frequently used throughout the article:

- A Reddit **Post** is text, link or an image posted by a registered user.
- **Comment** is a response to an active post on Reddit. It is a direct response to the post or a response to any comment made on a post, thus creating a nested structure of a tree graph with possibly any number of offsprings at any level.
- **Author** is a registered user on the platform who has at least one post or comment.
- **Score** is the difference between the number of upvotes and downvotes.

3.2 Definitions

We define the following quantities, which we will use in the rest of the article:

- **Age** is the time difference between the last comment on the post and creation of the post, measured in seconds (unless otherwise mentioned).
- **Effective Comments** are the total number of comments on a post made by users other than the author of the post.
- **Automoderator** Reddit's Official bot.
- **Deleted Author** Authors whose unique identity is absent while posts or comments are present in the metadata.

3.3 Data

We use two separate data sets of Reddit [15], in order to see if the data exhibits any qualitative changes along with the quantitative changes, over a gap of few years:

- Period I: 1 January 2008–31 December 2008,
- Period II: 1 August 2014–31 July 2015.

We considered posts and comments for entire periods of 1 year each along with associated meta-data like the title of a post, time of post/comment, subreddit topic, parent post/comment id, etc. We have considered only those comments which were made on a post within the periods of study. We also neglected the comments made during Period I to posts created before Period I. The same was followed while analyzing content for Period II as well. Table 1 summarizes the basic statistics of the data. It is important to note that after September 2015, there was a change in the number of fields that were being provided by Reddit API. So, in order to maintain consistency, we have used the data till July 2015. We analyzed the data using parallel computation on a Hadoop setup.

Table 1 2008 data table

	Period I	Period II
Number of posts	2,523,761	63,118,764
Posts with deleted authors	425,770 (16.87%)	12,346,042 (19.56%)
Posts with zero comments	1,536,962	23,417,869
Posts with one comment	591,489	9,011,332
Number of comments	7,242,871	613,385,507
Number of comments on posts of the period	7,224,539	608,654,680
Number of disconnected posts	219 (0.009%)	1380 (0.002%)
Number of removed comments	355 (0.004 %)	248,493 (0.04%)

Table 2 Used data variables for posts and comments

Posts	Comments
Author	Author
Created utc	Created utc
	Link id
Name	Name
Number of comments	Parent id

Table 2 shows set of variables from the available metadata for posts and comments that are used for our study. We have not used score in our analyses, except for the case of cyborgs.

4 Analysis of Post Evolution Patterns

To analyze the evolution of the posts, we calculate the age and number of comments for each post.

4.1 Mayfly Buzz

The probability density function (PDF) of the ages of all posts (Fig. 1) has a most likely value at 6 s for Period I, whereas the equivalent peak is smeared across values less than 6 s for Period II. Also, we observe that there is a shift in slope around the age of 1 day, after which, the PDF declines more quickly, suggesting that more posts tend to become inactive after a day. In fact, 88.6% of posts die within a day in period I and 71.1% of posts in period II. We call this post behavior *Mayfly Buzz*, which resonates with the idea of creating a day buzz. As seen in other social networking platforms, activity usually dies after a very short period of time. It is interesting to see a similar behavior on Reddit, which is a discussion platform as opposed to a microblogging site such as Twitter [11], where a post's average age is longer than a tweet.

4.2 Cyborg-Like Behavior

Figure 2 shows the age distribution (frequency) of all the posts which have a single comment. In Period I, there is a very prominent peak at 6 s, as found earlier (Fig. 1). It can be seen that the ages of 72.78% of these posts do not exceed 600 s (= 10 min). Period II looks very similar except the peak is seen at 5 s with an additional peak at 1 s. Furthermore, we analyzed posts whose first comment is posted within 6 s, which constitutes 43,138 posts for Period I and 1,804,374 for Period II. Out of these posts

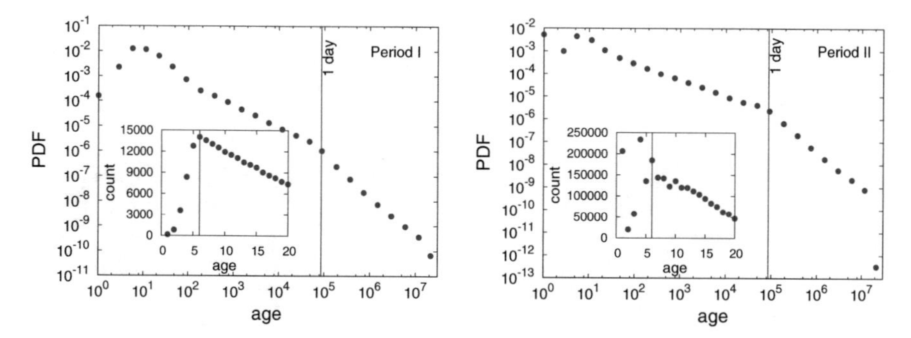

Fig. 1 The PDF of the age of a post (seconds) over the entire range of age for Periods I and II in double logarithmic scale. The nature of the probability distribution shows a prominent change of slope around ≈1 day, indicating that a large fraction of posts become inactive after that time. The insets show the corresponding histograms of the age distribution at small values of age, in double linear scale. Prominent peak around 6 s for Period I and at values less than that for Period II are notable

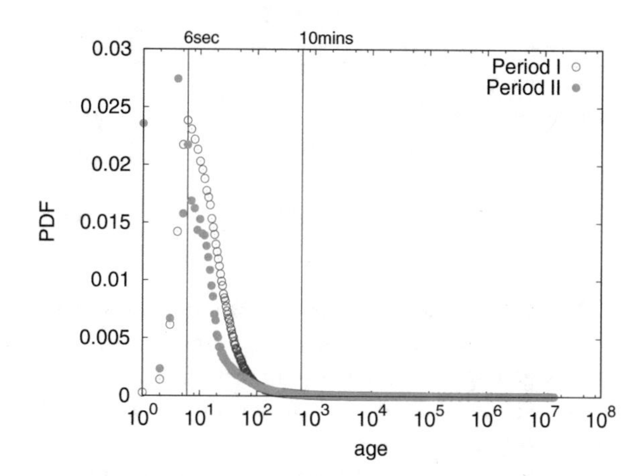

Fig. 2 PDF of ages for posts with one comment for Periods I and II

we found that approximately 17% and 20% posts have their first comment by the author of the post for Period I and Period II respectively. To understand this behavior of posting comment by the same user, we checked the number of characters in the first comment of these posts. For instance, we find that 83.9% and 79% posts have number of characters more than 100 for Period I and II respectively. It is crucial to mention that we have left out posts which contain links to web-pages in this analysis, which may be copied from a certain source and pasted in the posts. Choice of such figures are based on the rationale that writing such long comments within 6 s is quite impossible for a genuine human, and more likely to be done using automated means. Hence, we categorize these posts to be exhibiting a *cyborg-like* behavior, where these posts may be just an advertisement or a message that these users intend to propagate.

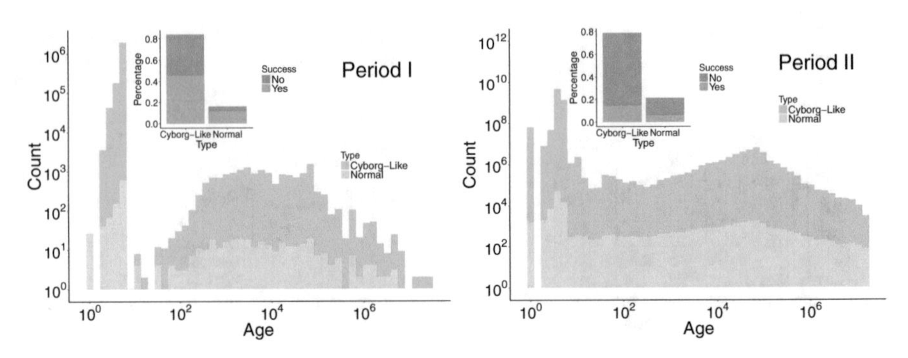

Fig. 3 The histograms for the actual age of the posts for which the first comment arrived within 6 s. The insets show distribution of posts over success rate

Table 3 Cyborg-like posts statistics

	Period I	Period II
Posts with first comment in less than 6 s	43,138	1,804,374
Posts with same author of first comment	7615	492,928
Cyborg-like posts	6389	387,845
Successful Cyborg-like posts	3446	70,237
Successful Non Cyborg-like posts	866	28,892
Unsuccessful Cyborg-like posts	2943	317,608
Unsuccessful Non Cyborg-like posts	360	76,191

We also define a 'success criterion' in order to check whether this type of posts were successful in garnering attention or responses from other users—if a post is getting any reaction (comment or vote) from other Reddit users, then they are considered 'successful' in drawing attention. We find that 53.93% and 18% *cyborg-like* posts were successful in Periods I and II respectively. While 70.63% of normal posts (which have comments with less than 100 characters) of Period I are successful, which can be assumed to be possibly done by humans (insets of Fig. 3), which comes to about 27.5% for Period II. Table 3 summarizes the data for this analysis. Hence, for Period I, we infer that machine generated content is less likely to garner interest as compared to human generated content. It is reasonable to assume the low success of the cyborg-like posts are due to the fact that lengthy comments and promotions/advertisements provide less room for any discussions. For Period II the cyborg-like posts have smaller success rate, which requires a further granular analysis, due to the increased richness in the variety of behavior in the cyborg-like posts in recent times.

4.3 Analysis of Depth and Breadth of a Post

Discussion happening on a post can be seen to have a tree-like structure. Depth of a post is defined by the maximum length of nested replies on a post and breadth of a post is defined as the maximum number of comments at a particular level. Figure 4 shows the variation of depth against the breadth of the posts for both the periods, with the heat map depicting the density. We observe that depth and breadth can grow independent of each other. This is most prominent in Period II, where posts with simultaneously large values of depth and breadth are rare or absent. The plot also shows that breadth grows more easily compared to depth, which can be attributed to the larger effort needed to grow or continue a nested discussion (increase depth) than to diversify a discussion by adding new parallel threads (increase breadth).

The probability distributions of depth shows fat tails for both periods. Additionally, there are prominent peaks at certain specific values 676 and 1000 which account for several counting threads in counting subreddits where users incrementally count successive alphabet pairs (AA, AB,...) and numbers up to 1000 by replying to one another, and subsequently start a new thread when the series is complete. The probability distribution of breadth also shows fat tails with power law decay for the very high values. The extreme values correspond to subreddits named 'Millionaire' where each user starts a new discussion at first level and thus contribute to increasing the breadth by unity.

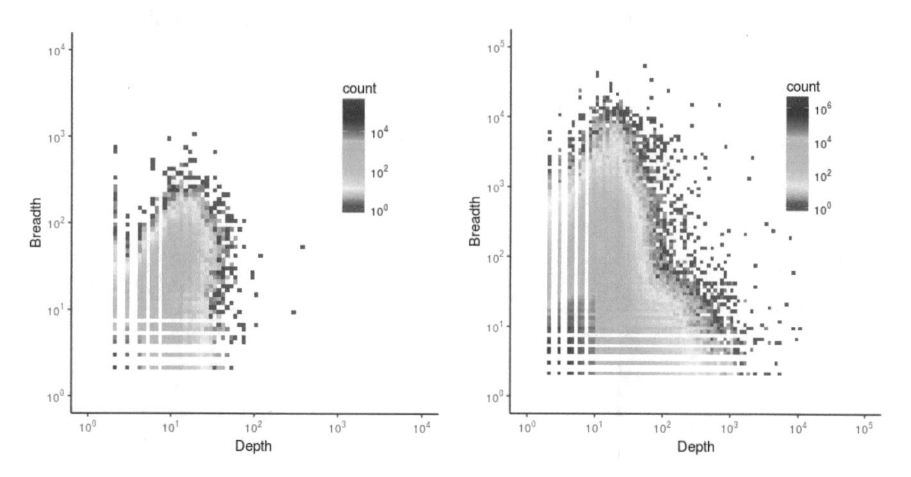

Fig. 4 Plot showing the depth of the posts against the breadth of the posts, with the density of posts shown through the heat map

4.4 Dynamics of Posts

One can easily characterize a temporal sequence of events by computing the time difference τ between successive events. Let us assume a sequence of events n, separated by a temporal distance $\tau_i, i = 1, 2, \ldots, n-1$. One can define the burstiness of a signal as [9]

$$B = \frac{\sigma - \mu}{\sigma + \mu} \quad (1)$$

where μ and σ are the mean and standard deviation of a sequence of τ. By definition, $B = 1$ for the most bursty signal, $B = 0$ for a neutral signal and $B = -1$ for a completely regular sequence. We compute a sequence of τ and subsequently the value of the burstiness B. This can be done for each author in terms of posts and comments, which corresponds to an author's posting burstiness or commenting burstiness. Additionally we can compute the burstiness of a post by considering all comments arriving in it.

Figure 5 shows the histogram of burstiness B for authors' posting behavior (authors with at least 100 posts), authors' commenting behavior (authors with at least 500 comments), and commenting behavior on posts (posts with at least 500 comments) for Period II. We find that this distribution is skew with a larger fraction of posts with positive burstiness value ($B > 0$), and in fact, the mean value $\tilde{B}$ is 0.31 for authors' posting, 0.45 for authors' commenting and 0.85 for all comments

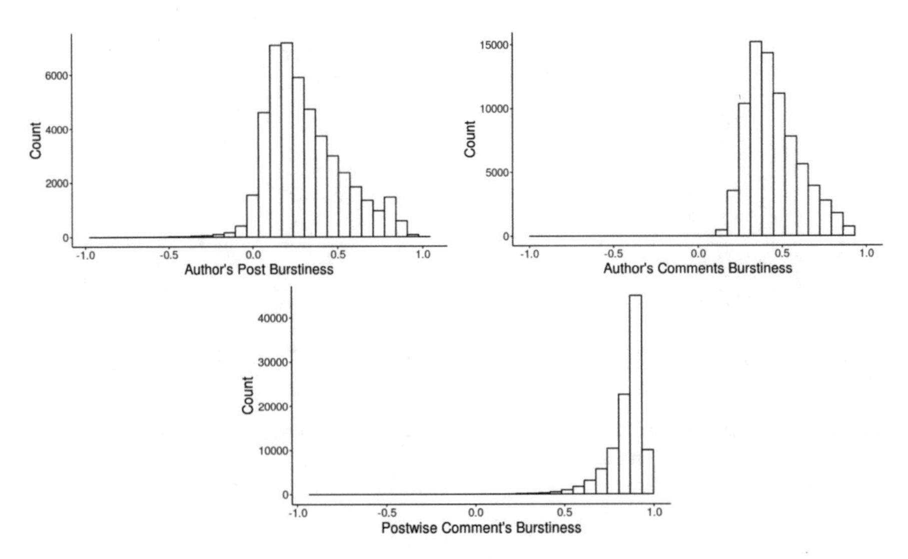

Fig. 5 Plot showing the histogram of burstiness B for (1) authors' posting behavior (authors with at least 100 posts), (2) authors' commenting behavior (authors with at least 500 comments), and (3) commenting behavior on posts (posts with at least 500 comments) for Period II

on posts. The first two quantities measure burstiness at the level of individuals and have seemingly similar distributions of burstiness and its mean, but the burstiness at the level of posts in terms of comments from various authors shows extremely skewed distribution of burstiness B with very large average value as well. This indicates that while activities at individual user level can be less bursty, the collective attention to a particular post can drive extremely bursty behavior. This reaffirms the known hypothesis that human communication show bursty patterns [9].

In real world communications, it is usual to find that the temporal spacing between successive human activities are quite heterogeneous, and range from completely random (following a Poisson distribution) to comments arriving at bursts.

4.5 Popular Post Dynamics

To understand the age dynamics of the popular posts and infer their behavior, plot the time evolution of the posts which have more than 500 comments. Three distinct categories are prominent (Fig. 6):

- *Early bloomers* are rapidly growing posts, accumulating over 75% of their total comments within 1 day, creating the *Mayfly Buzz* as discussed earlier,
- *Steady posts* are characterized by steady activity throughout their lifespan.
- *late bloomers* are slowly growing posts, which get suddenly very active at a late stage (after 30 days).

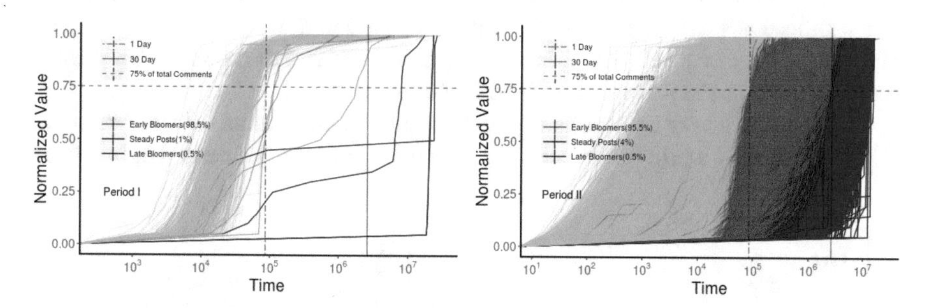

Fig. 6 The time evolution of the number of comments in a post (normalized by the final number of comments obtained within our defined time window of 1 year) for all posts having more than 500 comments. The data has been coarse-grained to aid visualization. The horizontal dotted line corresponds to 75% of the final number of comments. The time to reach this fraction is used to characterize the posts. The vertical lines are at 1 day and 30 days. Posts mostly active within 1 day garner 75% of their comments during that period (green). Some posts grow throughout their active life span taking time intermediate between 1 and 30 days to reach the 75% mark (red), while others grow slowly while becoming active at some later stage, beyond the 30 days period (blue). Plots are shown for Period I and Period II

We also study the evolution of the total number of comments with the age of each post, for all posts in our data. The overlaid binned average of all data indicates a marked departure in the gross behavior around 1 day.

5 Analysis of Interactions

The Reddit post-comment structure constitutes a tree graph, where the starting node is a post, and it can have its comments, and the comments can further garner replies. We defined a *limelight score* for each post based on the number of comments gathered as reply to a single first-level comment. In a way, this score can compute the depth of discussion around a single comment for a given post.

$$\text{Limelight Score} = \frac{\max(Comm_j)}{\sum_{k=1}^{N} Comm_j},$$

$Comm_j$ being the total number of comments under jth first level comment and N is the total number of first level comments for that post.

Figure 7 shows the histogram of the *Limelight scores* and the inset shows its CDF. We have considered posts that have at least 500 comments. We find that in Period I, 56% of the total posts contain one comment with *Limelight score* of at least 0.25, i.e., at least 25% of the discussion in this post is initiated and centered around a single comment. Similar behavior is also present in 31% of the posts in Period II. Additionally in Period II, a finite number of posts actually have Limelight score close to unity, indicating absolute dominance of one branch of the comment tree for those posts. We further observe that most of the time, the author of the post and that of the first level *Limelight* hogging comment are not the same. This holds for about 97% of the posts during Period I, for instance.

This leads to a very interesting insight that links virtual human behavior in social media to physical world social behavior. It is quite common scenario that in course

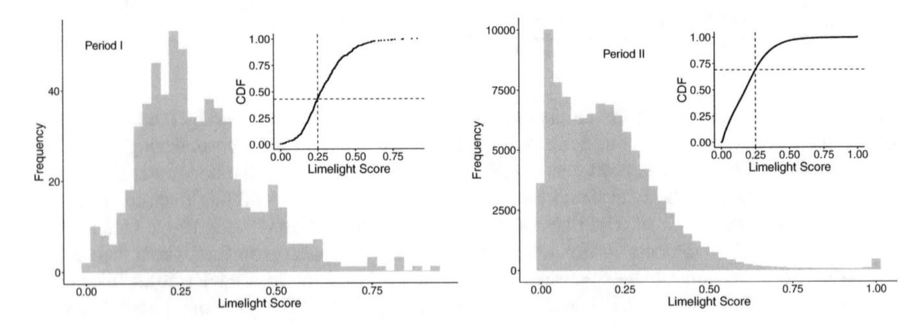

Fig. 7 Histograms for Limelight scores for all posts having at least 500 comments in Period I and Period II. The corresponding CDFs are shown in insets

of any group discussion, usually a few specific people apart from the presenter who pro-actively initiates a conversation by asking a question or making a comment, following which other people join the conversation by making comments or replies. Surprisingly, it is observed that lime-light hogging behavior is completely missing for posts whose authors exhibit *Cyborg-like* behavior. Thus, it can be inferred that posts automatically generated by bots have failed to garner human attention most of the times. However, rigorous studies can be undertaken in future to validate this inference.

To the best of our knowledge, characterizing content popularity by the depth of discussion around it has not been attempted before. Since it has been proved in earlier studies [8, 17] that the number of upvotes-downvotes are not meaningful indicators for measuring interestingness or popularity of content, we claim that this can be a good way to measure them.

6 Analysis of Author Behavior

To analyze author interactions, we started by defining a network with nodes representing unique authors and edges representing the interaction between the authors using comments. We define the in-degree and out-degree for each node based on the number of interactions, where a self loop is ignored. The gross statistics for the 3 categories of authors: (1) who only put up posts but don't comment on others' posts are the pure *content producers*, (2) who only comment are the pure *content consumers*, and (3) the rest of them indulge in both posting and commenting, are summarized in Table 4.

We try to quantify author interactions to assess their influence. If the total effective number of comments received is given by A and the total number of comments on others' posts is given by B, then we can define the **interaction score** of an author as $A/(A+B)$. This score is trivially zero for all authors who comment on others' posts but have not received any comments on their posts. Score is trivially 1, for an author who does not comment on others' posts but receives comments on one's own posts. This is rather rarely observed. For both periods, peaks at 0, 0.5 and 1 are prominent.

There are some distinct authors who have the ability to consistently garner a large number of comments on each of their posts. We analyze the average number

Table 4 Author table

	Period I	Period II
Total active authors	229,488	9,369,708
Total authors who only create posts	140,918	1,917,161
Total authors who only comments	39,764	3,019,676
Total authors who comment as well create posts	48,806	4,432,871

of effective comments received per post by authors, in order to quantify this. We observed that 22% of the authors in Period I and 6% in Period II have fewer effective comments than the number of posts that they have put up which corresponds no interaction for many posts. 11% in Period I and 13% in Period II have received one comment per post on the average. The rest, amounts to 67% for Period I and 81% for Period II received more comments than the number of posts put up for the respective periods.

The discussion above indicates that authors who receive more attention on their posts are seemingly the ones who comment on others' posts. Simply put, in order to gain attention on social media, authors need to be reciprocative in nature, which is also indicated by the peak at interaction score of 0.5. It also emphasizes the fact that the social media interactions are dominated by mutual gratification.

7 On Signatures of Controversies

Contrary to the graph based approaches common in the literature, which deal with controversies in social media platforms (see, e.g., Ref. [5]), we take a rather moderate, statistical approach, which attempts to lay down the foundation for a rigorous, sentiment analysis based approach in the future.

We observe that in the popular posts (with more than 500 comments), some comments have been deleted either by the author of the comment or by the moderators of the subreddits. In the latter case, deletion of a comment can happen only if the author made a comment that violates the rules of the subreddit set by the moderators, that can potentially lead to controversy in a social discussion platform. For further analysis, we have calculated the ratio of the number of deleted comments to the total number of comments, which can serve as a proxy for the measure of controversiality of a post and call it the *Controversiality Score*. In the top panel of Fig. 8 we plot the Controversiality Score of a post against the number of unique authors for Period II. We observe that the plot branches roughly into two components for lower number of unique authors, one each for very high and very low level of controversiality score. This branching is absent beyond a certain number of unique authors, which is roughly 200 for our case. In the top panel of Fig. 8, the colors map to the number of comments in the posts.

We observe that when there are fewer unique authors (less than 200 in our case) contributing in a discussion to a post, the outcomes can be quite extreme—it can either see a very high degree of controversy or a very low degree of controversy, as is seen by the left part of the graph. Beyond 200 unique authors, this extreme diversity vanishes—in fact, very high values of controversiality are absent. This indicates that more controversiality occurs within smaller groups than over larger ones.

We further wanted to check if controversiality is related to the popularity of the subreddit in which the post is created. We define the popularity of a subreddit as the total number of posts that are being created in that subreddit during the period, and

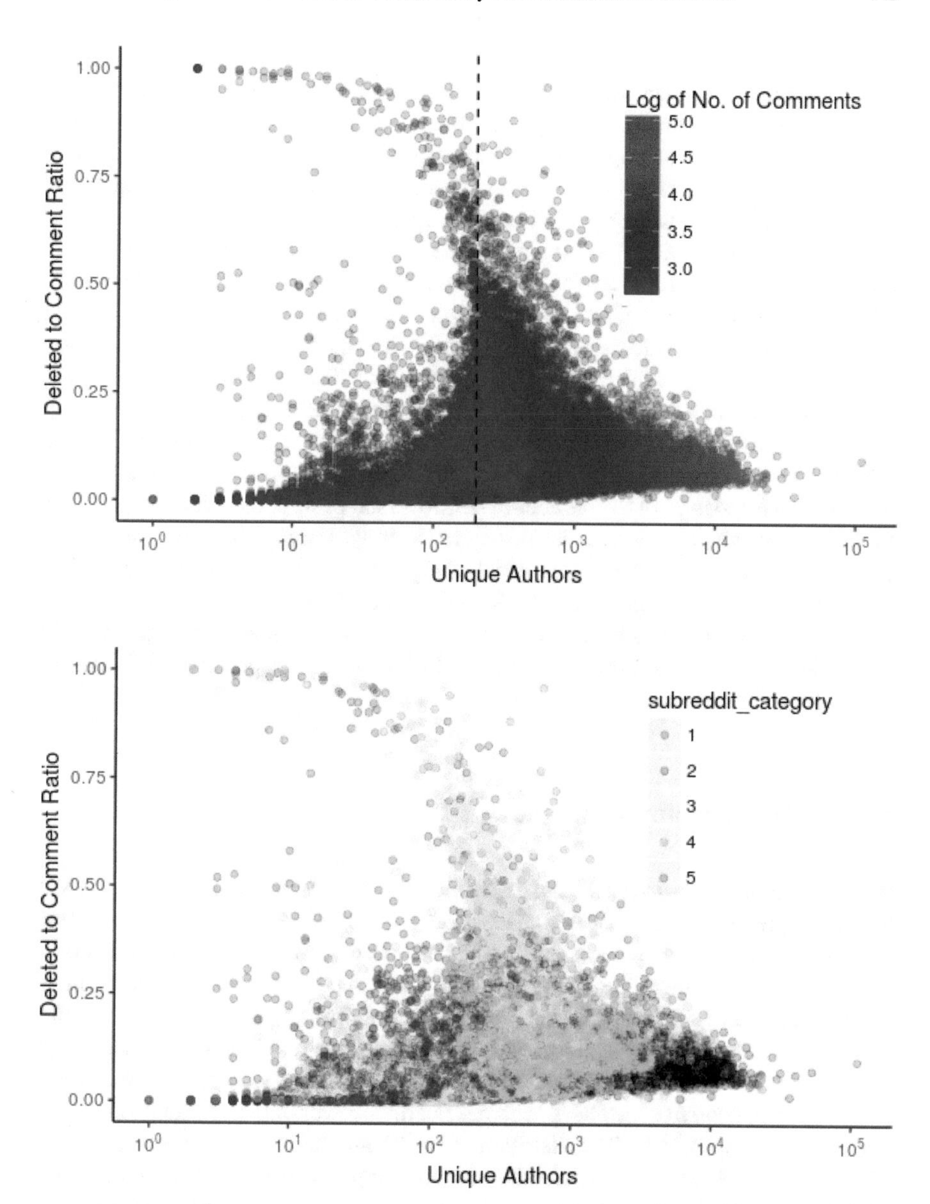

Fig. 8 (Top) The plot of controversiality score against the number of unique authors in the post, for all posts in Period II with at least 500 comments. Each post is coloured according to the number of comments it has. (Bottom) Same plot where the color of the point is according to the category of the subreddit in which they belong (discussion in text)

divide them into 5 categories, 1 being the least popular subreddit and 5 being the most popular subreddit (1: 1–10 posts, 2: 11–100, 3: 101–1000, 4: 1001–2000, 5: above 2000). In the bottom panel of Fig. 8, we plot the controversiality score of posts against the number of unique authors with colors indicating the popularity category of the subreddit. We can see that the popular subreddits (categories 4 and 5) have low controversiality score, around 0.25 or less, while high controversiality scores are prevalent in the less popular categories. Hence, we infer that controversiality is observed in smaller, closely knit groups (akin to *contempt breeds contempt*), and users stay clear of controversies in larger groups.

We have also checked the controversiality of the individual subreddits. We consider subreddits which have at least 100 posts. A post is considered *controversial* if its controversiality score is more than 0.2 i.e., more than 20% of comments of that posts are deleted. We calculated controversiality score of a subreddit as the fraction of posts in it that are controversial. The top panel of Fig. 9 shows the controversial score of those subreddits in which users have posted at least 100 posts.

To check which users are responsible for the above, we can compute author-wise controversiality score. We can extend the above definition to *author controversiality score* which is the ratio of the number of controversial posts to the total number posts by the author where controversial posts are those with more than 20% of deleted comments in the posts. The bottom panel of Fig. 9 shows the author controversiality score of authors who have more than 50 posts in our data.

The above measures are the indicators that tell us in which Reddit community (subreddit) controversial posts are put up and which user is responsible for initiating the controversy.

8 Conclusions and Outlook

A large, community-driven social network and discussion platform like Reddit harbors a plethora of behaviors for the users concerned. A huge fraction of posts are left uncommented, while some gather of a considerable amount of attention through actions on them like comments and votes. The distribution of the number of comments on posts show correlation through the power law tail [18], and the behavior of authors show a large variation—while many authors simultaneously create post and write comments, there are also a large fraction of purely *content producers* and *content consumers*, who restrict themselves only to either posting and commenting respectively. The distribution of the number of comments by unique authors exhibit lognormal distribution for the largest values, which indicates an underlying multiplicative process, and thus a strong correlation between authors. Each post stays active as comments are added and thus discussions are produced. However, a huge fraction of posts are left with only a single comment, and within them, a majority receive that only single comment within 6 s, containing a large number characters unlikely to be written by an average human, indicating a *Cyborg-like* behavior. Further to that, a large fraction of posts become inactive around the age of 1 day. This is consistent with the average active time of posts reported for micro-blogging site such as Twitter [11]. When we studied the time evolution of the top commented posts, we found three broad classes for the posts: (1) *early bloomers*

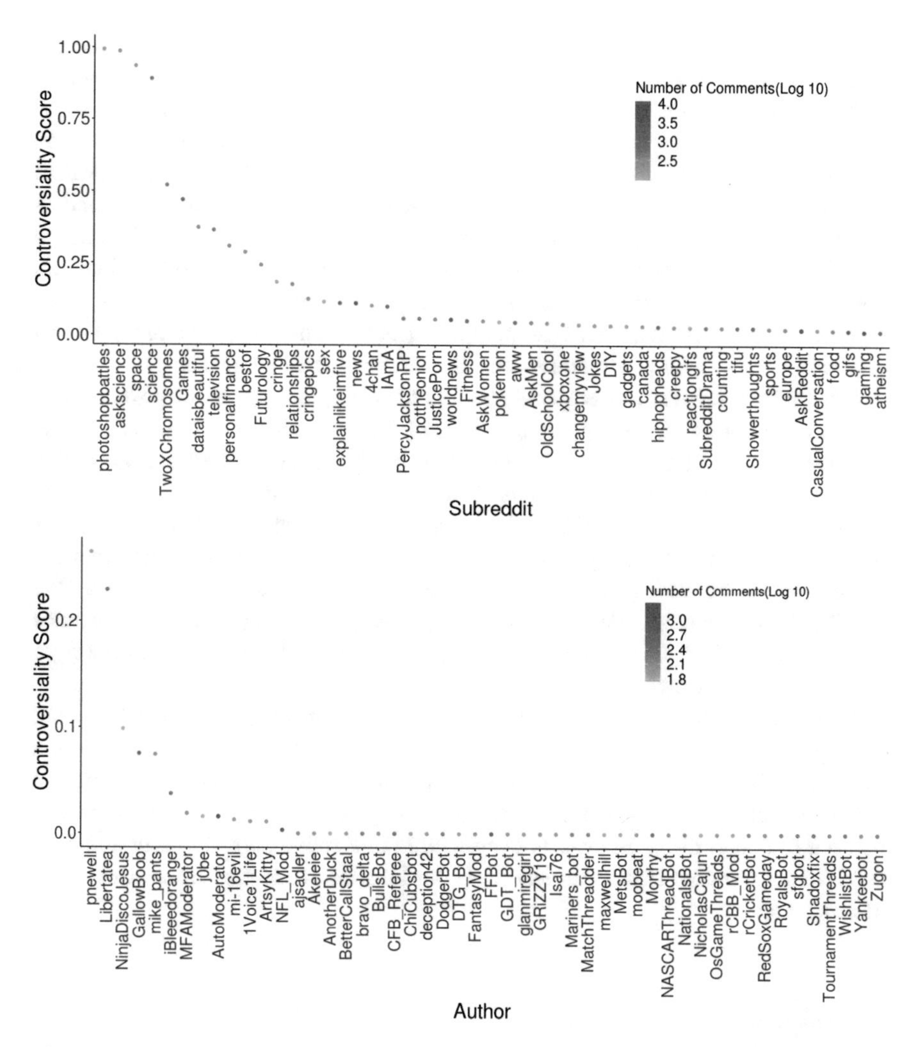

Fig. 9 (Top) Plot for the controversiality score for the subreddits which have more than 100 posts. (Bottom) Plot for the controversiality score for the authors which have more than 50 posts

who gather more than 75% of their lifetime comments within a day, (2) *steady posts* whose number of comments grow steadily throughout their lifespan, and (3) *late bloomers* who show very little activity until steadily gathering comments near the end of their lifespan. The early bloomers contribute to what we term as *Mayfly Buzz*, and constitute the majority of the posts. Posts also show *limelight hogging* behavior and we find that 56% of posts in Period I and 31% of posts in Period II have *limelight score* above 0.25, indicating that for a large fraction of posts, at least one-fourth of the total weight of the discussions are contributed by one chain of comments. In fact, this measure can be a more meaningful indicator of the interestingness or popularity of the content, compared to just votes or only number of comments. Social media discussion threads sometimes contain controversial content, and in Reddit this is moderated by deleting posts or comments. Our study tries to measure the controversiality from the fraction of such deletions, at the level of posts, authors and subreddits. We observe that controversiality is more prevalent in small, closely knit groups than large ones. Analysis of actual content can lead us to a better understanding, which we plan to carry out in future studies. Our initial measurements of sentiment on the text content of comments around deletions did not indicate any significant signal, but probably a further careful analysis of other quantities along with sentiment can help us formulate a unique indicator which can eventually be used for prediction/forecasting of unruly textual events.

We also investigated the temporal patterns of events, in terms of the posts created by individual authors, the comments created by individual authors, as well as the comments on popular posts. All of them show bursty behavior of events, validating the fact that human communication is usually bursty in nature.

With the increasing use of social media even within closed groups as well as organizations, understanding human behaviors and ability to characterize them is turning out to be an important task with potential impact and applications. One possible application of understanding temporal patterns of group behavior in such a scenario can be focused on injecting the right content or advertisement for the right group at the right time.

Our analysis brings out a variety of behavioral elements from the authors and through their interactions. There are few authors who are able to generate quite a lot of activity across a large number of posts. Going ahead, analysis of change of sentiment can turn out to be interesting. The insights gained from this analysis can be used to model different aspects from a large interactive population. In addition, predicting the recent trends can lead to better targeted reach e.g., innovative usage of *memes* etc.

References

1. C. Buntain, J. Golbeck, Identifying social roles in reddit using network structure, in *Proceedings of the 23rd International Conference on World Wide Web* (ACM, New York, 2014), pp. 615–620
2. Y. Choi, Y. Jung, S.H. Myaeng, Identifying controversial issues and their sub-topics in news articles, in *Pacific-Asia Workshop on Intelligence and Security Informatics* (Springer, Heidelberg, 2010), pp. 140–153
3. M.D. Conover, J. Ratkiewicz, M. Francisco, B. Gonçalves, F. Menczer, A. Flammini, Political polarization on twitter, in *Fifth International AAAI Conference on Weblogs and Social Media* (2011)
4. D. Gaffney, J.N. Matias, Caveat emptor, computational social science: large-scale missing data in a widely-published reddit corpus. PLoS One **13**(7), e0200162 (2018)
5. K. Garimella, G.D.F. Morales, A. Gionis, M. Mathioudakis, Quantifying controversy on social media. Assoc. Comput. Mach. Trans. Soc. Comput. **1**(1), 3 (2018)

6. M. Glenski, T. Weninger, Predicting user-interactions on reddit, in *Proceedings of the 2017 IEEE/ACM International Conference on Advances in Social Networks Analysis and Mining 2017* (ACM, New York, 2017), pp. 609–612
7. M. Glenski, T. Weninger, Rating effects on social news posts and comments. Assoc. Comput. Mach. Trans. Intell. Syst. Technol. **8**(6), 78 (2017)
8. M. Glenski, C. Pennycuff, T. Weninger, Consumers and curators: browsing and voting patterns on reddit. IEEE Trans. Comput. Soc. Syst. **4**(4), 196–206 (2017)
9. K.I. Goh, A.L. Barabási, Burstiness and memory in complex systems. Europhys. Lett. **81**(4), 48002 (2008)
10. V. Gómez, A. Kaltenbrunner, V. López, Statistical analysis of the social network and discussion threads in slashdot, in *Proceeding 17th International Conference on World Wide Web* (ACM, New York, 2008), pp. 645–654
11. H. Kwak, C. Lee, H. Park, S. Moon, What is twitter, a social network or a news media? in *Proceeding 19th International Conference on World Wide Web* (ACM, New York, 2010), pp. 591–600
12. D. Lazer, A. Pentland, L. Adamic, S. Aral, A.L. Barabási, D. Brewer, N. Christakis, N. Contractor, J. Fowler, M. Gutmann, T. Jebara, G. King, M. Macy, D. Roy, M. Van Alstyne, Computational social science. Science **323**(5915), 721–723 (2009)
13. Y. Mejova, A.X. Zhang, N. Diakopoulos, C. Castillo, *Controversy and Sentiment in Online News* (2014). arXiv preprint: 1409.8152
14. R. Mills, Researching social new—is Reddit.com a mouthpiece for the 'Hive Mind', or a collective intelligence approach to information overload? in *Proceedings of the Twelfth International Conference, the Social Impact of Social Computing ETHICOMP 2011* (Sheffield Hallam University, Sheffield, 2011), pp. 300–310
15. pushshift.io: databases. https://pushshift.io/resources/databases/ (2017). Accessed 01 Oct 2017
16. P. Singer, F. Flöck, C. Meinhart, E. Zeitfogel, M. Strohmaier, Evolution of reddit: from the front page of the internet to a self-referential community? in *Proceedings of the 23rd International Conference on World Wide Web* (ACM, New York, 2014), pp. 517–522
17. G. Stoddard, Popularity and quality in social news aggregators: a study of reddit and hacker news, in *Proceedings of the 24th International Conference on World Wide Web* (ACM, New York, 2015), pp. 815–818
18. S. Thukral, H. Meisheri, T. Kataria, A. Agarwal, I. Verma, A. Chatterjee, L. Dey, Analyzing behavioral trends in community driven discussion platforms like reddit, in *2018 IEEE/ACM International Conference on Advances in Social Networks Analysis and Mining (ASONAM)* (IEEE, Piscataway, 2018), pp. 662–669
19. S. Wasserman, K. Faust, *Social Network Analysis: Methods and Applications*, vol. 8 (Cambridge University, Cambridge, 1994)
20. T. Weninger, X.A. Zhu, J. Han, An exploration of discussion threads in social news sites: a case study of the Reddit community, in *2013 IEEE/ACM International Conference on Advances in Social Networks Analysis and Mining (ASONAM)* (IEEE, Piscataway, 2013), pp. 579–583
21. T. Yano, N.A. Smith, What's worthy of comment? content and comment volume in political blogs, in *ICWSM* (2010)

Mining Habitual User Choices from Google Maps History Logs

Iraklis Varlamis, Christos Sardianos, and Grigoris Bouras

1 Introduction

The popularity of smartphones, their voice assistants and navigators made them a habitual choice for drivers, who take advantage of its hands-free nature in order to easily receive driving directions in a constant basis. This explains why 80% of the Internet users own a smartphone [1], while mobile application usage is growing by 6% every year. In the same time, modern smart devices are equipped with multiple ambient sensors that provide great amounts of data, which can be analyzed to discover useful information such as daily user patterns, trajectory patterns etc. According to Business Insider[1] as of September 2018, 13% of consumers use Siri and 6% use Google Assistant through Apple CarPlay and Android Auto respectively and 20% of consumers use voice in their car through their smartphone via Bluetooth. Navigation applications constantly collect user location information in order to optimize route suggestions and provide up-to-date traffic notifications. The collection and analysis of geo-location information collected by navigation applications can provide surprising insights about habitual user behaviors, which can then be utilized for personalizing user experience and provide useful information to each user. Together with the advent of Location-Based Social networks [2] that assume users to share location information with other users for social purposes, the rise in the use of navigation applications set a new challenge, which is the identification of habitual user choices for places and routes [3].

[1] https://www.businessinsider.com/voice-assistants-car-vs-smartphones-2018-11.

I. Varlamis (✉) · C. Sardianos · G. Bouras
Department of Informatics and Telematics, Harokopio University of Athens, Athens, Greece
e-mail: varlamis@hua.gr; sardianos@hua.gr; it21239@hua.gr

M. Kaya et al. (eds.), *Putting Social Media and Networking Data in Practice for Education, Planning, Prediction and Recommendation*, Lecture Notes in Social Networks, https://doi.org/10.1007/978-3-030-33698-1_9

The information about the places that a user has visited can be exploited in many ways, for example, for promoting POIs that match user's profile [4], for recommending alternatives to his/her fellow users and friends in a location-based social network [5], for extracting useful statistics about the popularity of POIs, or even for creating package recommendations for the visitors of a city [6]. A recommender system can take advantage of user location history only, can add content information from social networks, from explicit and implicit user preferences, from third-part services that provide information about POIs (e.g. Yelp, Foursquare, Open Street Maps etc.), or even can take advantage of the temporal information behind each check-in [7].

This work, extends our previous work on the analysis of Google History location data [8], which extracted user stay-points and trajectories among them, abstracted the information about frequently visited places to preferred amenity types and generalized trajectory information to user preferred transportation type by place. Among the contributions of this work is that it provides an end-to-end implementation of a framework for mining user habits from trajectory data. It also extends existing semantic trajectory enrichment frameworks that extract stop and move information from trajectories and enhance stops and moves with semantics with an additional information abstraction layer, in which user habits are extracted from repetitive user trajectories.

In this work, we consider both types of information that can be extracted from user location data, namely location and trajectory information, but mainly focus on user frequent trajectories. More specifically, a new methodology for generating route alternatives is introduced, which is based on trajectory segmentation and sub-trajectory clustering. The proposed methodology employs

- trajectory partitioning techniques that use both stay-points and turn-points as the points where major changes happen in a user's route,
- trajectory clustering techniques that define sub-trajectory segment similarity using segments' start and end points and
- a trajectory re-construction technique that joins sub-trajectories to re-create the original route and its alternatives.

In Sect. 2 that follows, we summarize the most important works that relate to the extraction of user habits from user location data and take advantage of this information for generating personalized recommendations. In Sect. 3 we briefly discuss the main concepts of this paper, which are frequently visited points, frequently traversed routes and trajectories and in Sect. 4 we explain how we have extended our methodology to be allow the composition of route alternatives for frequent user routes. Section 5 provides an overview of the implementation details of the system and Sect. 6 shows some results that demonstrate the methodology and the information it can extract from a user's Google Maps history logs. The same section illustrates the frequent trajectories and the re-composed route alternatives that have been extracted using the new methodology. Finally, Sect. 7 summarizes our progress so far and explains how a recommendation system can take advantage of the extracted habitual knowledge in order to deliver the right recommendation at the right moment.

2 Related Work

The concept of mining useful knowledge from spatio-temporal data has been discussed several times in the related literature. The survey work of Zheng [9] in trajectory data mining, summarizes all paradigms of trajectory mining and the issues that must be considered. More specifically, when it comes to user trajectories, early works [10] analyze GPS logs in order to mine interesting locations and travel sequences, employ user location history to measure user similarity [11], or identify and assign significance to semantic locations based on GPS records [12], whereas more recent works [13] take advantage of user location history and a richer content (i.e. user sentiment, user interest and location properties) in order to better match users to locations and recommend POIs of interest to users.

When the focus is on discovering interesting places in user trajectories, then clustering algorithms are employed. Density-based algorithms, such as DBSCAN and its spatio-temporal variations [14] are used in the points that compose many trajectories in order to detect dense clusters of points that corresponds to repetitive user presence in an area. When the focus is on finding interesting trajectories, then it is important to partition largest trajectories to sub-trajectories that connect user POIs and then to define a trajectory similarity measure [15] before applying a clustering algorithm [16]. Based on the generated trajectory (or sub-trajectory) clusters it is possible to plan and recommend sightseeing tours that are popular among the visitors of a region [17].

Collaborative applications have boosted the interest of GPS log mining and introduced location and activity recommender systems and location-based social networks. The survey work of Bao et al. [18] provides a taxonomy of recommender systems that build around social and location based information for supporting Location-Based social networks. Such applications, are based on user location and trajectory data [19–21], the overall user behavior and social circle in order to recommend POIs [22] or trajectories [23] that connect places of interest.

Despite the long interest on user location and trajectory history, there have only been a few works that take advantage of user's latent behavior patterns in order to provide personalized recommendations. It is worth to mention the work of He et al. [22], who attempt to jointly model next POI recommendation under the influence of user's latent behavior pattern. Authors adopt a third-rank tensor to model the successive check-in behaviors of a user in POIs and fuse it to the personalized Markov chain of observed successive user check-ins, in order to improve POI recommendations. They generalize user check-in history at day of week level in the time dimension and at POI category in the POI dimension. However, in their model they limit user profiling only on the POIs and not on the trajectories between POIs.

In a slightly different context, authors in [7] introduce the concept of temporal matching between user profile and POI popularity. They profile the temporal pattern of area activity around POIs, using information from taxi pick-ups and drop-offs and propose that every user has a latent daily-repeated personalized temporal regularity, which decides when he/she is likely to explore POIs every day.

The method proposed in our previous work [8] was based on the analysis and abstraction of user geo-location data in order to extract useful information concerning the user's behavior and habits. This work focuses on frequently traversed user routes and combines methods and techniques from the related literature in order to propose a new methodology for generating route alternatives. The methodology is based on the partitioning of trajectories to smaller sub-trajectories, which are first clustered and then reconnected in order to re-compose the user route and possible alternative routes. The proposed approach can be used by navigation applications and their recommendation algorithms in order to evaluate traffic conditions and propose alternative routes to users, that match their previous preferences and consequently are easier for the users to adopt. The proposed approach differs from existing systems in that it extracts user habits in the form of temporal patterns that repeatedly occur in user logs and uses this habitual information for creating a better user experience for navigation and recommendation systems.

3 Extraction of User Preferences

The first step for finding user preferred locations and routes is the detection of user stay points, which is followed by the detection of frequently traversed user trajectories between consecutive stay points. User habits can be enriched by semantically annotating stay points with type or amenities information and trajectories with the preferred transportation method.

3.1 Extraction of Stay Points

A *stay point* SP stands for a geographic region, where a user stays over a certain time interval [10]. The extraction of a stay point depends on two parameters, a time interval threshold (T_{threh}) and a distance threshold (D_{threh}). For example, for the set of GPS points depicted in Fig. 1, that correspond to the consecutive positions of a moving user, the stay point SP can be regarded as the virtual location containing a subset of consecutive GPS points $SP = \{p_m, p_{m+1}, \ldots, p_n\}$, where $\forall i, j \in [m, n]$ it holds that $Distance(p_i, p_j) \leq D_{threh}$ and $|p_i.T - p_j.T| \geq T_{threh}$, where $p_i.T$ is the timestamp associated with point p_i. The centroid of all GPS points that grouped

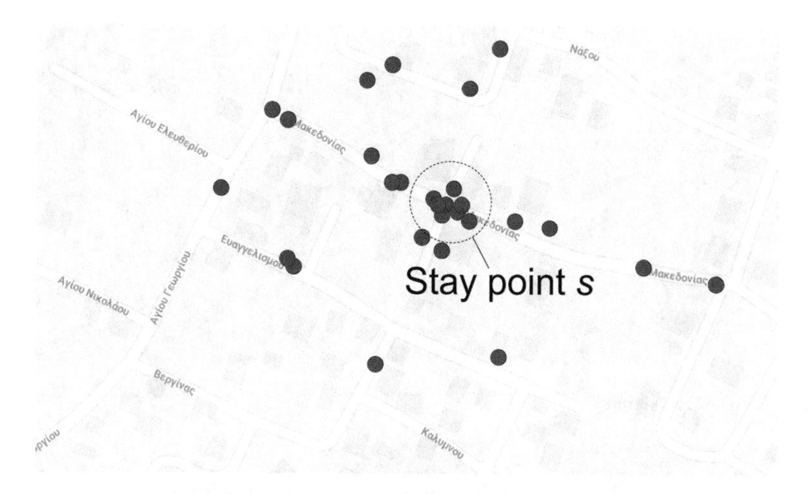

Fig. 1 Stay point example

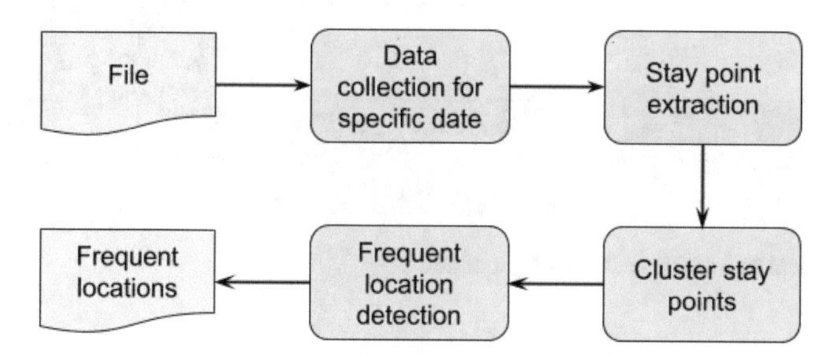

Fig. 2 Pipeline for extracting frequent user locations

under the same stay point location, is used as the GPS coordinates of stay point SP and a maximum radius for the cluster is also kept with the stay point. Finally, since the stay point represents the stay of the user in a location for a time period, we keep the start and end time-stamp with the stay point. So, a stay point is characterized as: $SP =< lat, lon, radius, t_{start}, t_{end} >$.

The steps of the stay point extraction process are depicted in Fig. 2.

3.2 Extraction of Frequent User Trajectories

A *user trajectory* Tr_{ij} is defined as the sequence of GPS points between two stay points SP_i and SP_j. Tr_{ij} defines a route comprising a series of GPS locations in chronological order. Stay points denote the end of one trajectory and the beginning of another. The timestamp difference between two consecutive points (GPS locations' timestamps) in a trajectory, is exceeded, so that they cannot be considered to belong in the same stay point. Figure 3 presents an example of

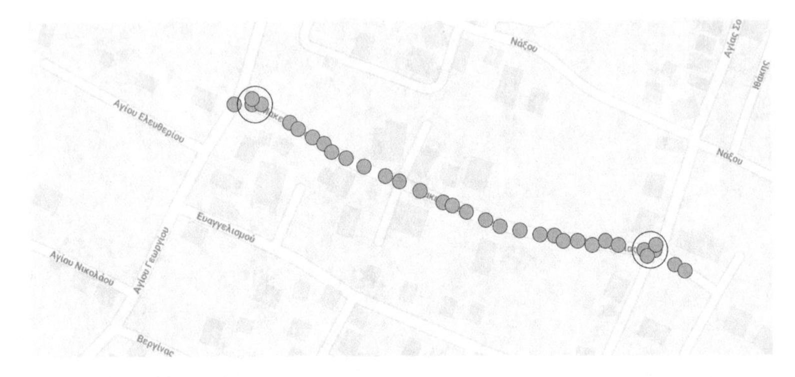

Fig. 3 Trajectory example

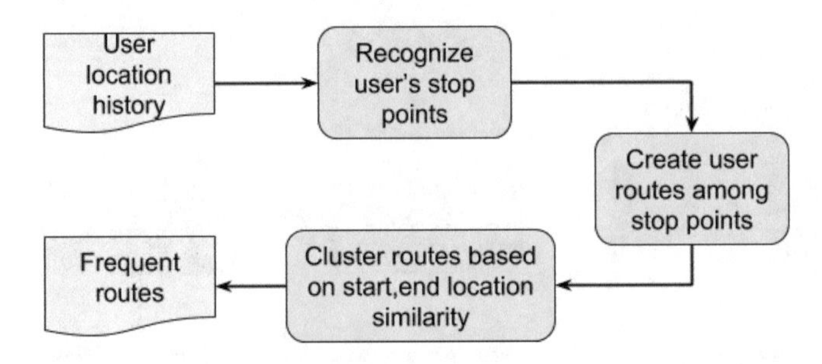

Fig. 4 Frequent trajectories extraction pipeline

a user trajectory connecting two consecutive stay points. So a user trajectory is characterized by a set of GPS coordinates a start and an end time-stamp as: $Tr = < \{(lat_i, lon_i)\}, tr_{start}, tr_{end} >$.

Frequent routes aggregate repeated user trajectories, which not necessarily completely match, but have the same start and end location. So, frequent trajectories are characterized by the start and end location a set of start (departure) and end (arrival) time-stamps and type of movement for each trajectory instance, as follows:

$$FreqTr = < POI_{start}, POI_{end}, \{(mov_{type}, t_{start}, t_{end})_i\} >$$

The processing pipeline is summarized in Fig. 4.

3.3 *Frequent Route Summarization*

Frequent routes are associated with points that a user visits frequently, even on a daily basis. The analysis of such routes may uncover habitual user routes, but also may highlight temporal variations that depend on several factors, such as emerging user needs or varying user preferences. A typical example can be the route from home to work and back, which is performed on an almost daily basis, but which

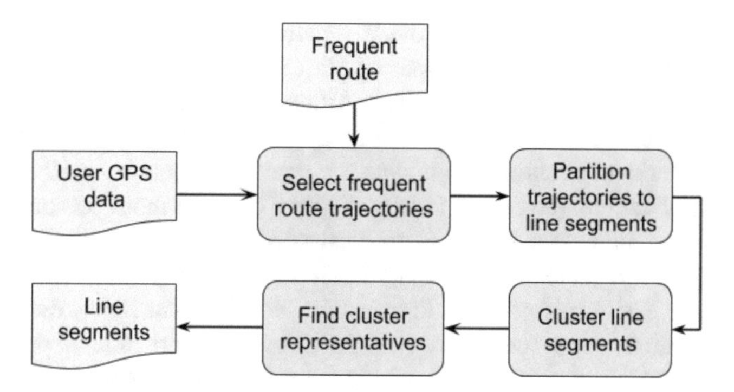

Fig. 5 The pipeline for summarizing frequent route's trajectories

may contain several intermediate stops, e.g. for shopping, for picking up the kids from school, etc., and several deviations or route alternatives, that depend on traffic or weather conditions, or on an emerging need for an additional stop or deviation. In order to find the places visited by a user on a frequently traversed route and the alternative paths followed during a frequent route, we apply a segment and cluster methodology on the daily trajectories that relate to the specific route.

According to the proposed methodology, each trajectory $Tr_x = < \{(lat_i, lon_i)\}, tr_{begin}, tr_{end} >$ that connects the starting point $(lat_{begin}, lon_{begin})$ with the end point (lat_{end}, lon_{end}) (e.g. start is home and end is work) is first partitioned to smaller sub-trajectories Tr_{x_j} using certain points along the route that declare a repeated major change, such as stay points and turn points. Stay points declare a user stop for several seconds (e.g. a traffic light, a bus stop or an indented stop in a POI, whereas turn points declare a major change in user's movement direction such as a more than 30° direction change (at a junction or round-about). The first step of the proposed methodology finds the partitioning points from all the alternative route implementations and then partitions the trajectory to sub-trajectories.

All the sub-trajectory segments extracted in the previous step are clustered to groups of similar segments, using a trajectory similarity measure and a trajectory clustering algorithm. Each sub-trajectory cluster represents a part of the frequently repeated user route, that connects two intermediate points of the route. The processing pipeline is summarized in Fig. 5.

3.4 Semantic Enhancement of Stay Points and User Trajectories

When a stay point or a trajectory is extracted, the proposed method applies a semantic enrichment process, which attaches additional information that can be useful for extracting user habits at a later stage. The information for characterizing stay points can be provided from different sources. However, the main sources are

POI information services, which provide semantic annotations for popular points of interest as well as for common types of POIs (e.g. leisure places, sport facilities, public buildings, transportation hubs etc.). Although POI information services cover a large amount of locations, there are still stay points that can not be characterized without the user's intervention. Following the practice of popular on-the-go driving directions applications (e.g. Google Maps), users can add more personal semantic information for their own points of interest such as "Home" or "Work". These explicitly provided annotations usually refer to frequently accessed stay points that cannot be easily mapped to a known POI. A stay point SP is defined as any geographic region where the user stayed for at least a short time of period (e.g. a store or an office) or even a location where the user slightly moved around, but not away from it (e.g. a park or a stadium etc.). So after extracting the possible stay points of the user, we access POI information services and search for nearby POIs, at a short range and always within the stay point limits. Using the same services, we are able to characterize the stay point by the type, such as: Park, Stadium, Leisure ground, Athletic center, Beach, Shopping center, Cafeteria etc. At the end of this step a stay point is characterized as:

$$SP =< lat, lon, POI_{type}, POI_{category}, t_{start}, t_{end} >$$

The semantic annotation of user trajectories, mainly refers to the application of data mining techniques to the trajectory points information (latitude, longitude and timestamp) and the detection of user type of movement across the trajectory. This allows to detect at a later stage, the preferred way of movement for specific trajectories or overall, whether the user uses public or private means of transportation etc. We treat the problem of detection of user movement type across a trajectory as classification problem [24, 25], and build on our previous work on the topic [26, 27]. The movement categories can be: motionless, walking, running, riding a bike, driving a car, being on the bus, being on a train/metro, and any other type of movement. At the end of this step a trajectory is characterized as:

$$Tr =< \{(lat_i, lon_i)\}, mov_{type}, POI_{start}, POI_{end}, t_{start}, t_{end} >$$

3.5 Abstracting User Data to User Habits

The term user habits describes a routine of behavior that the user subconsciously repeats regularly [28, 29]. In the case of geo-location data habits are repetitive user activities such as: being in the same stay point at the same time or day of a week, taking the same trajectory at the same time or week day. In this routine, users' take habitual decisions that slightly modify the daily plans, without changing the overall aim of the routine. For example, people choose among alternative routes when commuting depending on whether there is traffic or not, whether it is raining or sunny etc.

The analysis of user trajectories for a long time, allows to detect the places that a user frequently visits and the routes that the user traverses more often on week days or weekends, in the morning or afternoon [30]. At user level, the trajectories can define a set of user behaviors, highlight user habits and allow recommendation algorithms or similar applications to provide user-tailored *location and time based* recommendations. Finally, in a collaborative environment (e.g. in location based social networks) users can be compared based on their detected habits and collaborative filtering can be fused with user location and time context (and habits) information.

The extraction of user habits begins with the detection of frequent occurring patterns and continues with the abstraction of information at various levels of granularity in time (e.g. time zones, days, etc.), POI type, movement type or route. The result of this process in terms of user routes comprises frequently traversed segments that compose the main route and its alternatives, for one or more users that move in the same or similar routes. The overall approach for analysis, can be described as a three-tier analysis with multiple tasks in each tier. Figure 6 describes the basic architecture of our application.

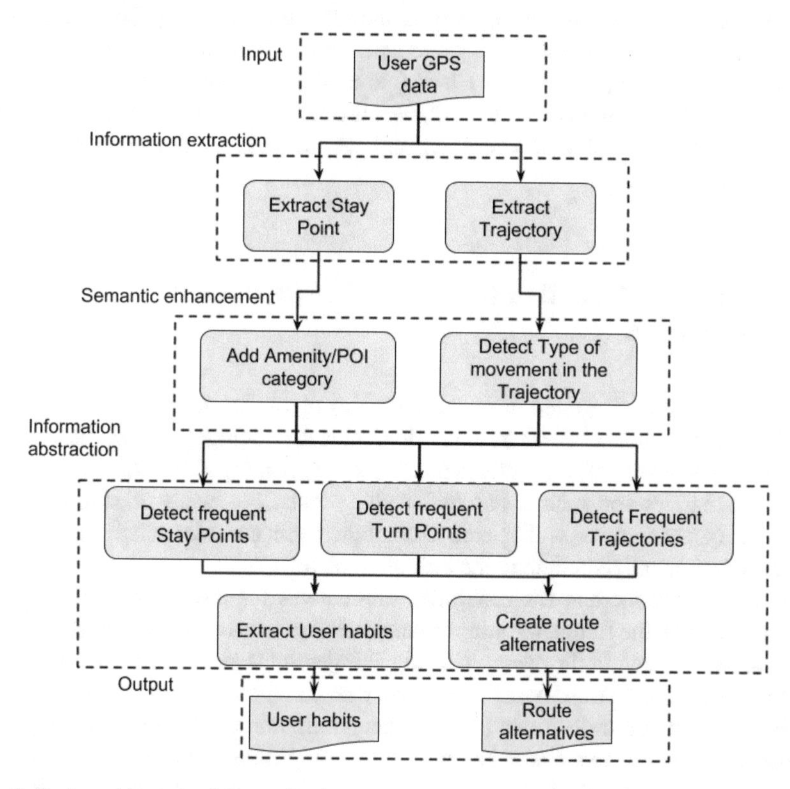

Fig. 6 Basic architecture of the application

Each layer in the proposed architecture implements algorithms that are widely applicable in similar problems. First, in the information extraction layer, the detection of stay points and user trajectories between consecutive stay points adopts the methodology initially proposed in the SMOT (Stops and Moves of Trajectories) algorithm [31], which defines *stops* as polygons with a minimum stay duration and *moves* as contiguous sub-trajectories between two temporally consecutive stops. Using a density based clustering algorithm such as DB-Scan is a straightforward choice since we are interested in clusters with many points at a small spatio-temporal distance. The same algorithm has been used by many researchers in the field, since 2008, that it was introduced for trajectory points clustering [14].

Second, in the information enhancement layer two different strategies are applied: (1) the labeling with POI or amenity type information, (2) the labeling of move sub-trajectories with the type of movement. In the former case, we implement a typical engineering approach that takes advantage of online POI databases as it has been done in related works [32, 33], whereas in the latter, we capitalized in our previous work in the field [27]. The output of these two layers is a rich trajectory database that follows the principles of the CONSTAnT data model [34].

Finally, in the information abstraction layer, we implement the step of behavioral knowledge extraction from trajectories, as initially defined in [35]. Even very recent works in the field [36] employ discrete state representation of the trajectory and probabilistic models (e.g. Markov Chains) to model user movement and stay events and discover daily patterns, in our work, we introduce additional context parameters, such as time-of-day, day-of-week etc. and we are able to generated more fine-grained user habits.

4 Creating Alternative Routes for Frequent User Trajectories

During a frequently repeated movement, such as that from home to work, it is quite common that the user chooses an alternative route depending on several factors such as traffic, weather conditions, planned activities etc. An algorithm that recommends route alternatives for such a habitual activity must be aware of historical user decisions (route choices) and be able to combine them, consider actual user position and destination and recommend the best alternative.

The proposed method for generating alternatives for frequently traversed user routes, builds on the frequent route summarization process described in the previous section. As depicted in the example of Fig. 7 trajectories are first partitioned to sub-trajectories, based on the detection of points where major changes happen (e.g. stops or turns). Then sub-trajectories (line segments) are clustered using a line segment similarity measure and a clustering algorithm. The proposed methodology for route summarization is based on the partition-and-group framework introduced in [16], which is expanded with a trajectory partitioning step (pre-processing) and a route to

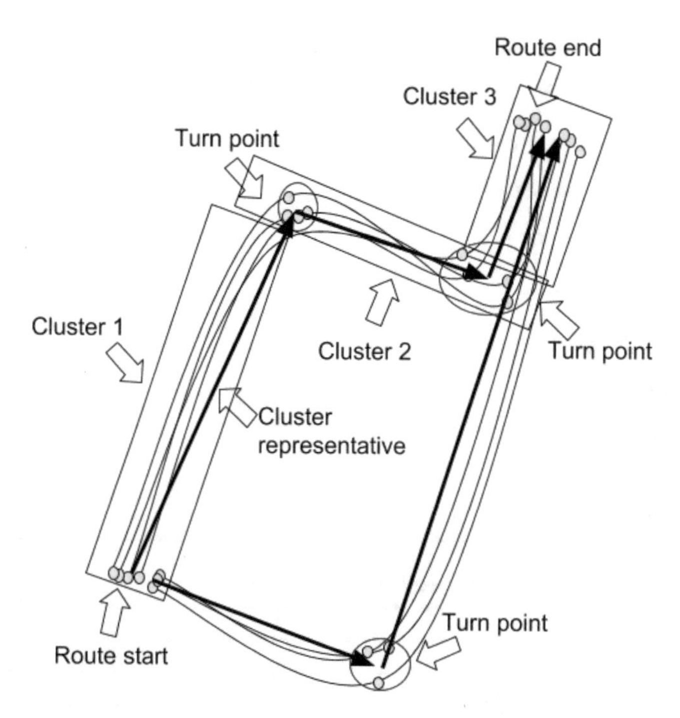

Fig. 7 An example of partitioning and clustering of multiple trajectories

graph representation step (post-processing) that results with to a directed weighted graph for each frequent route.

A sketch of the modified frequent route summarization algorithm is given in Algorithm 1. Table 1 summarizes the notation used in our algorithms.

In order to compose route alternatives for a frequent user route, we employ the information extracted by the Frequent Route Summarization algorithm, which is represented by a graph, where the stay and turn points are the vertices and the representative segments are the edges. In order to reconstruct the route alternatives, it is necessary to connect the representative sub-trajectories of clusters in a way that they form the user route. For example, the three alternative routes of Fig. 8, marked with different colors are the result of connecting all the clusters of points and denote the three different ways chosen by a user when driving from home (bottom left) to work (top right).

The alternative route composition is performed by running a k-shortest path algorithm in the generated graph, which returns the top-k paths that connect the starting point to the end point in decreasing edge weight sum. The edge weight can be subject to their popularity, the average travelling time, the length of the respective segment etc. So, if the weight of the edge is the distance to travel, then the shortest paths will be ordered by shortest distance, if the weight is relative to the time needed to cross an edge then the paths will be ordered by shortest time and

Algorithm 1 The frequent route summarization algorithm

Input: A set of trajectories $I = TR_1; \ldots; TR_n$
Output: (1) A set of representative segments $RI = RS_1; \ldots; RS_m$ (2) A set of stay/turn point clusters $STP = p_1; \ldots; p_n$

```
for each TR ∈ I do
    S = splittosegments(TR)
    for each s ∈ S do
        AllSegments.add(s)
    end
end
O = cluster(AllSegments)
for each C ∈ O do
    RS = getrepresentativesegment(C)
    RI.add(RS)
    STP.add(RS_start)
    STP.add(RS_end)
end
```

Table 1 A summary of the notation used in the algorithms

Notation	Description
TR	Trajectory
RS	Representative segment
RI	Set of representative segments
STP	Stay point cluster set
p	A single stay point cluster
S	Set of trajectory segments
s	A single segment
C	A cluster of very similar trajectory segments
AR	Alternative route
G	A graph containing the main user stay points as vertices (V) and the trajectories between them as edges (E)

so on. Algorithm 2 provides a sketch of the process that generates alternatives for a frequent user route.

Using the case of Fig. 8 as an example, the graph representation will comprise a set of vertices $V = \{HOME, A, B, \ldots, I, WORK\}$ and a set of edges $E = \{(HOME, A), (A, E), (E, F) \ldots\}$. Example route alternatives will be $\{HOME, A, E, F, I, WORK\}$, $\{HOME, B, D, G, H, I, WORK\}$ and so on. Depending on the actual position of the user different route alternatives are ranked higher. For example, if the user is still at home there exist 5 route alternatives to commute, whereas if the user is commuting and is already at point D, then the only route alternatives will be $\{D, G, H, I, WORK\}$ and $\{D, F, I, WORK\}$.

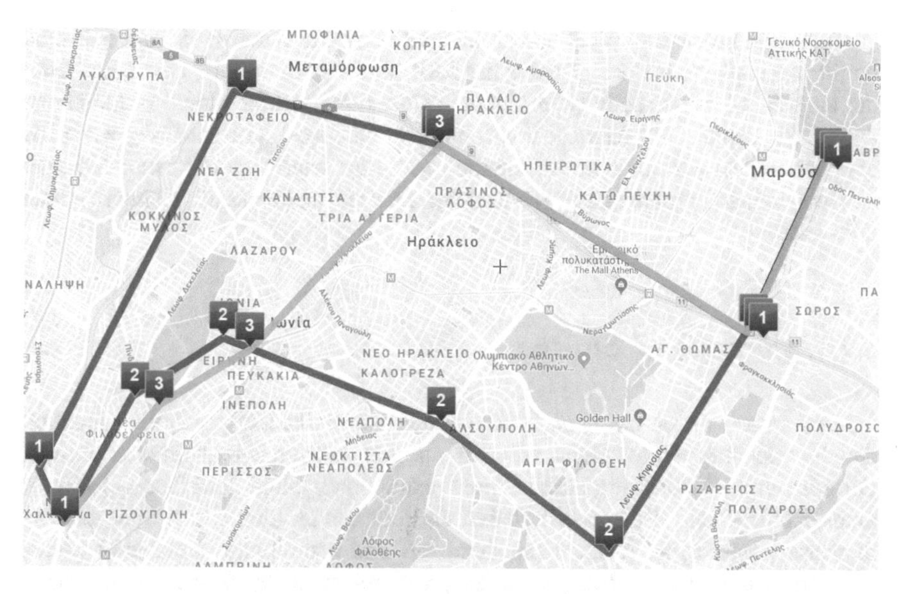

Fig. 8 An example of alternative routes for commuting

Algorithm 2 The algorithm for creating alternative routes for a frequent user route

Input: A set of trajectories $I = TR_1; \ldots; TR_n$
Output: A set of alternative routes $AR = AR_1; \ldots; AR_n$
$RI, STP = FrequentRouteSummarize(I)$
for *each* $RS \in RI$ **do**
  $V = V \cup RS_{begin}$
  $V = V \cup RS_{end}$
  $E = E \cup (RS_{begin}, RS_{end}, weight)$
end
$G = G(V, E)$
$AR = getkshortestpaths(G, k)$

5 System Implementation

5.1 Data Collection

The first step of the proposed method is the collection of user GPS data. Although in previous works we employed actual GPS data collected by the user smart-phone, in this work we use data from Google Maps history, which are imported as KML (Keyhole Markup Language) files to the system. KML is a file format based on the XML standard and uses a tag-based structure with nested elements and attributes to express geographic data (such as locations, image overlays, video links and modeling information like lines, shapes, 3D images and points) in location browsers such as Google Earth and Google Maps.

The same processing pipeline can be applied to the actual GPS data collected by the smart-phone instead of using the KML file. The information extracted from the analysis of a user activity in a certain time-frame can be stored in the phone and all the actual GPS data for this frame can be erased. For example, when the user commutes, we can store information about the trajectory (e.g. start/end time and location, and probably a few intermediate points) and erase all the intermediate GPS data.

5.2 Information Extraction

(1) Extraction of User Stay Points The initial task on the analysis of user location data is to identify the locations, where the user stays for a certain amount of time. For this purpose, following the visit point extraction method described in [37], we employ DBSCAN [38], a density-based clustering algorithm, which finds clusters of dense points using a range threshold *eps* and a minimum number of points $MinPts$ within this range as parameters. We implement a spatio-temporal version of DBSCAN (the distance of two points is a linear combination of geographic distance and time distance), which clusters together neighboring (in space and time) GPS traces and ignores all other points (considers them noise). Depending on the frequency of recorded GPS spots, the distance threshold of interest and a moving speed threshold, we can compute an acceptable value for $MinPts$. The parameterization of the algorithm has been described in [27].

A spatial clustering of user location data, using a density-based clustering algorithm, is expected to identify areas with very dense recorded spots and areas where the user passed through at a quick pace. This clustering will provide information for places that user spends time during the day, even when the user is standing or walking around (in a park) or running (in a stadium). The use of time distance in the distance measure of DBSCAN will change the resulting clusters, and will allow to distinguish between a stay point and a point that the user crosses several times but in different (distant) timestamps. A time-ignorant DBSCAN will detect a single cluster for all points, whereas a time-aware version will detect separate clusters. An incremental version of DBSCAN, allows to cluster the most recent GPS traces of a user and detect the stay points as they occur, and consequently assign all intermediate points to the trajectory.

Figure 9 depicts the results of the stay point extraction process on a map. On the left part all the GPS points, before the detection of stay points are shown on the map in red, whereas on the right part the detected stay points only are marked with green color.

(2) Detection of User Trajectories The detection of user trajectories is binded to the detection of stay points and trajectories are directly defined as the sets of GPS tracks between two consecutive stay points. Considering the fact that a large set of stay points may exist in user's GPS logs, a respectively large set of trajectories are

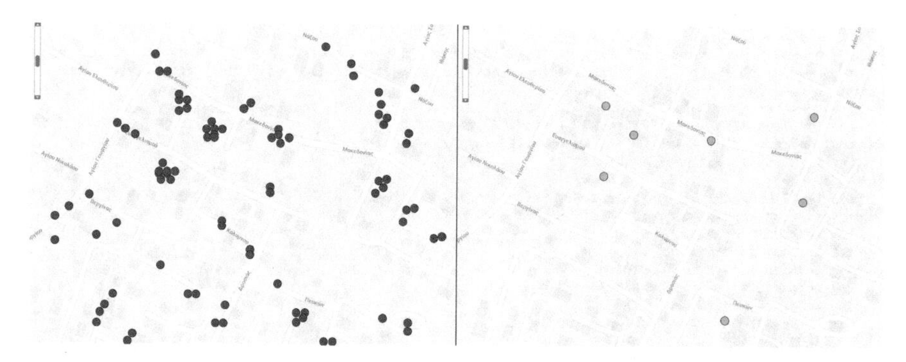

Fig. 9 All GPS points of user (on the left) and the results of detected stay points (on the right)

formed among the different stay points. So the result of this information extraction step comprises two sets: a set of user's stay points and a set of trajectories that join consecutive stay points.

5.3 *Information Enhancement*

(1) Semantic Characterization of Stay Points After user's stay points and trajectories are detected, they are annotated with additional information, which is collected by third party services or extracted by data mining algorithms, and which can be used in the user habits extraction step. The characterization of stay points step employs the OpenStreetMaps service, which offers an API for retrieving information about various POIs in a geographical area. For this purpose, a bounding box is created for each stay point, using the GPS coordinates of the stay point as the bounding box center and a range that does not exceed the radius of the respective cluster. The OpenStreetMaps API is accessed to retrieve POIs within the geographical area defined by the bounding box and it responds with an XML formatted result, as depicted in Fig. 10.

The file contains all possible Points-of-Interest marked with tags that characterize the type of POIs inside the bounding box of the stay point, such as:

- amenity
- public transport
- shop
- sport
- leisure

From the locations returned by the POI service, the closest to the stay point is used to characterize the stay point. The result of this process is that the user stayed at a specific POI (which is of certain type and category) for a specific time period.

```xml
<?xml version="1.0" encoding="UTF-8"?>
<osm version="0.6" generator="CGImap 0.4.0 (22136 thorn-03.openstreetmap.org)" copyright="OpenStreetMap
contributors" attribution="http://www.openstreetmap.org/copyright" license="http://opendatacommons.
org/licenses/odbl/1-0/">
 <bounds minlat="38.0340500" minlon="23.7369800" maxlat="38.0346400" maxlon="23.7375200"/>
 <node id="519954080" visible="true" version="3" changeset="10846515" timestamp="2012-03-02T10:12:51Z"
 user="armitatz" uid="414661" lat="38.0357216" lon="23.7355228">
  <tag k="highway" v="traffic_signals"/>
 </node>
 <node id="2750852492" visible="true" version="1" changeset="21374126" timestamp="2014-03-28T22:34:57Z"
 user="Chris Makridis" uid="1227858" lat="38.0341596" lon="23.7374317">
  <tag k="addr:housenumber" v="4"/>
  <tag k="addr:postcode" v="14341"/>
  <tag k="addr:street" v="Βρυούλων"/>
  <tag k="amenity" v="restaurant"/>
  <tag k="name" v="Εν Αιθρία"/>
  <tag k="website" v="www.enaithria.com"/>
 </node>
```


Fig. 10 Sample of the XML response with tags for stay points

(2) Detecting Type of Movement in a Trajectory For the semantic annotation of a user trajectory, we process all consecutive GPS traces in order to detect user movement speed, user direction and user speed changes, we also check if the traces are near a public transportation (PT) stop or on a known PT route. Building on our previous work on the topic [26, 27] we classify each trace individually and then classify the trajectory as a whole. If there exists a set of pre-classified movement samples we train a personalized model for each user, else a pre-trained generic model is used, which uses a set of direct (latitude, longitude, timestamp) and indirect features (speed, speed changes, distance from transportation related POIs such as bus stops or metro stations) in order to characterize how the user moved across a trajectory [27]. Since our original classifier is incremental and annotates the last trajectory part with the detected movement type it is frequent that for a long trajectory, more than one movement types have occurred (e.g. the user drives but stops at the traffic lights or is stuck behind a bus for part of the trajectory). When the next stay point is detected and the trajectory is completed, a post-processing step aggregates this information for all segments of the trajectory and assigns the movement type that most likely matches to the specific trajectory.

To provide an example, let's assume the daily commute of a user as depicted in Fig. 11. The user drives from home to the nearest train station, parks the car and takes the train from station A to the nearest station (station B) at work, then walks to reach the work place. The stay point detection algorithm will detect four stay points (home, parking lot of train station A, train station B, work) and three trajectories that connect them. The first trajectory will be annotated as driving, the second as moving by metro/train and the last as walking. The sub-trajectory from the train station's parking lot to the station building will not be detected if the two places are close to each other. The walk of the user from the parking lot to the train station will also be considered as part of user's visit to the specific stay point (i.e. train station A).

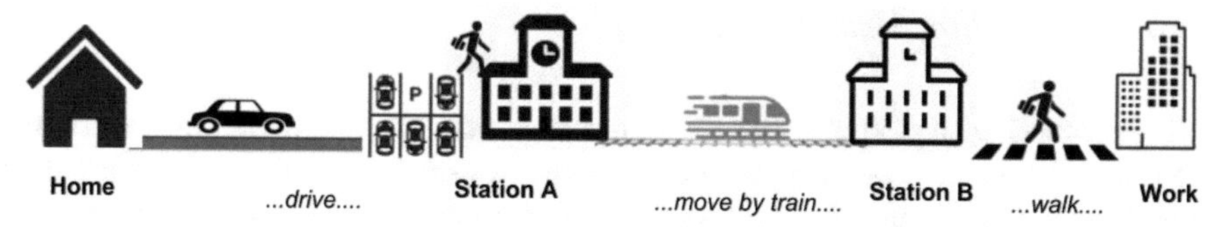

Fig. 11 An example of a user going to work. The detection and annotation of stay points and trajectories

5.4 *Information Abstraction*

(1) Detection of Frequent User Stay Points The next phase of the process is to extract the frequent stay points and trajectories of a user based on his/her location history. By analyzing the stay points of a user for a long time period, we can find if the user tends to visit specific stay points more frequently. To achieve this, we cluster stay points using a distance-based only DBSCAN and result with clusters containing points that have been visited many times by the user. We rank clusters in descending order of size and keep the top ranked stay points for a user. Along with the GPS coordinates of each stay point, we have time-stamp information concerning the start and end time of the user's stay. A first step of information abstraction is to find the preferred days or time zones for a user to visit a stay point. The result of this step can be similar to the following: *The user has visited train station A n times this month. The preferred days are week days, and the preferred time zones are early in the morning and the afternoon.*

(2) Detection of Frequent Trajectories Similar to the analysis of stay points, the analysis of user trajectories will highlight the preferred movement paths of the user and the preferred way of movement. We focus only on the frequent stay points for a user and considering the set of trajectories that the user has followed to go from one stay point to another. We apply the clustering-based sequential mining (CBM) algorithm [39] over the set of trajectories and the output of this process is the set of most frequent trajectories followed by the user. In detail, the CBM algorithm is based on the clustering of the set of points that belong to the trajectory, so in our case the input of the algorithm is the set of user trajectories that connect user frequently accessed stay points and two parameters, namely s and ξ. Parameter s defines the square area occupied by each cluster on the map and parameter ξ defines the minimum number of points that a cluster has to contain in order to be considered as active (that means that is frequently part of the user's trajectories).

5.5 Extracting User Habits

This processing step aims to use the information extracted from the previous steps and find frequently occurring activity patterns in user logs, which will define the user's habits. The analysis of information concerning frequently visited locations, which the user visits periodically, and frequent trajectories that the user follows to reach these destinations, will discover user's tendencies on going to a specific place (e.g. home → work → gym → restaurant → home) at specific dates and times.

The process of user habits extraction, can be treated as an association rules extraction problem. Before extracting frequent patters and interesting rules, it is important to process the annotated stay point and trajectory information and get different levels of abstraction. For example, different places that have been visited by the user and have been annotated as restaurants, bars or cafeterias can be generalized to the category amenity and lead to rules with stronger support, the time-stamp information can be mapped to day zones (e.g. morning, afternoon, evening) or days, using different levels of granularity. The output of this information abstraction step is fed to the association rule extraction algorithm, which in our case is the Apriori algorithm [40].

5.6 Creating Alternative Routes

As explained in the previous section, the proposed method for generating alternatives for frequently traversed user routes, follows a step-wise approach. For the partitioning of trajectories, both the stay and turn points have been employed. For this purpose, we use is the GPS and timestamp information from user GPS logs and the TrajLib library[2] to process this basic information and extract information about the trajectory at each point, such as bearing, moving speed and acceleration [41].

After the trajectories have been partitioned to segments, all segments are clustered using the distance function introduced in [16], which combines three components (1) the perpendicular distance ($d_{\perp}$), (2) the parallel distance ($d_{\parallel}$), and (3) the angle distance (d_{θ}) as illustrated in Fig. 12. The k-means algorithm has been used to group similar line-segments together.

The representative line segment for each trajectory cluster is created by connecting the centroid of the starting points (of all segments in the cluster) with the centroid of the end points. The start and end centroids are mapped to graph vertices and the cluster is mapped to a weighted and directed edge.

In the step of alternative route composition, the weighted directed graph is given as input to a k-shortest path extraction algorithm. More specifically, an implementation of the Yen's algorithm [42] for computing single-source K-shortest

[2]https://github.com/metemaad/TrajLib.

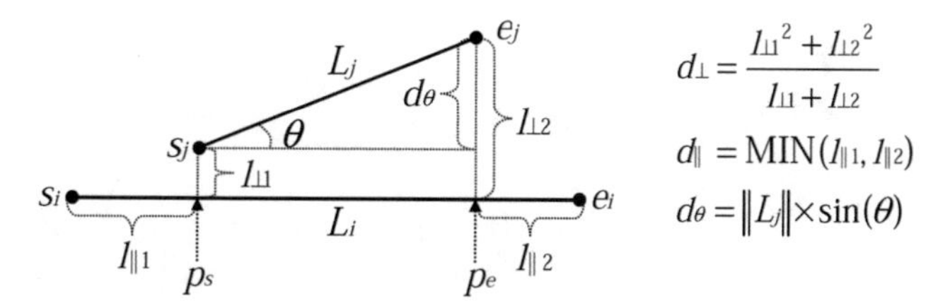

Fig. 12 The distance between two sub-trajectory line segments

loopless paths for a graph with non-negative edge weights has been used.[3] The result of this process is a list of alternative routes for a specific frequent route. This last step can be repeated every time the user moves along the frequent route, using the actual user position as source and the same destination (route destination) in order to re-evaluate the k-shortest paths.

6 Real Case Demonstration

In order to demonstrate our proposed method for extracting user habits from user trajectory data, we developed an application that takes Google Maps History files (KML files) as input and processes them following the process described in the previous sections. The application is written in Java and is available as a standalone Java program (Fig. 13).

Through the application the user has a set of options that can trigger the analysis of the location data contained in the KML file, which may span several days or months. The user can either analyze the loaded data and depict them over an OpenStreetMap map embedded in the user interface, or extract the results of the information extraction and enhancement step to output files for further analysis. Using these files as input, we can extract the user's frequent stay points and trajectories, which can be displayed over an OpenStreetMap layer or exported to separate files.

The analysis of user stay points leads to a set of user's tendencies like the ones displayed in Fig. 14.

These files are then fed to the Apriori algorithm to extract user habits. The input data that we use consist of the user's extracted stay points combined with the time-stamp of occurrence after converting the actual start and end time-stamps into daytime zone (e.g. morning, afternoon, night) and day (e.g. weekday, weekend), the type of movement at that moment and/or the type of the amenity. The set of categorical features are fed into the Apriori association rules algorithm. So, when the Apriori algorithm is fed with information in the form of:

[3]https://github.com/Pent00/YenKSP.

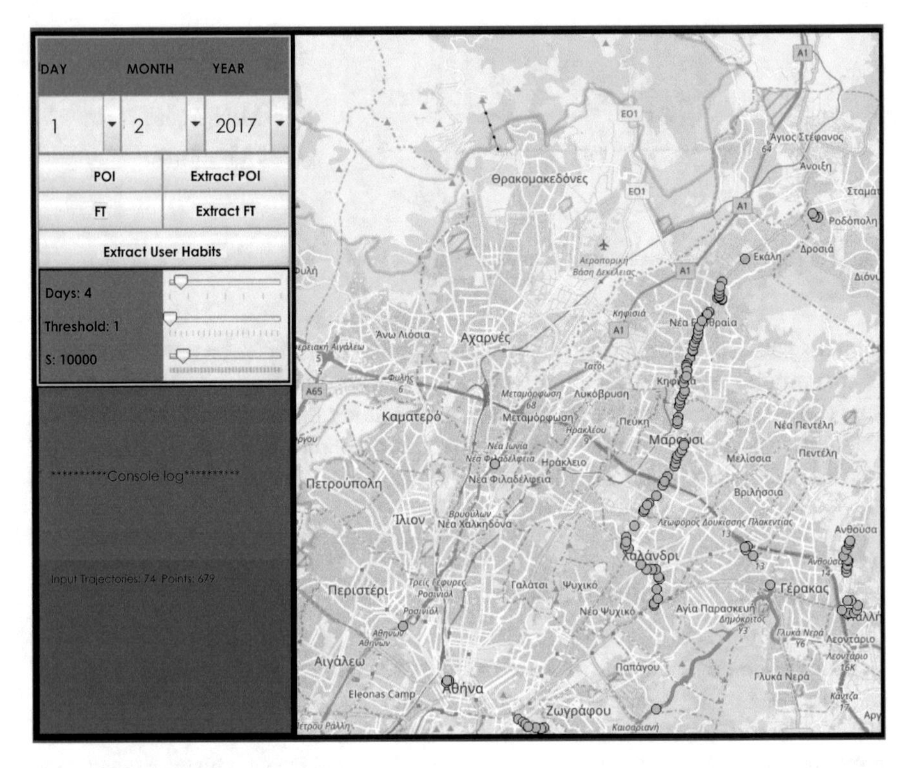

Fig. 13 Snapshot of the application

```
Latitude        Longitude       Time
37.8883883      23.9400781      Sun Jan 01 18:11:51 EET 2017 - Sun Jan 01 22:58:04 EET 2017

Location frequence:45
Frequent Day: Sunday

Latitude        Longitude       Time
37.9033496      23.7499417      Sun Jan 08 20:00:25 EET 2017 - Sun Jan 08 22:51:04 EET 2017

Location frequence:40
Frequent Day: Sunday

Latitude        Longitude       Time
37.8608787      23.7534746      Sun Jan 01 23:35:14 EET 2017 - Mon Jan 02 00:35:33 EET 2017

Location frequence:37
Frequent Day: Monday

Place name:Barón
Amenity:cafe    ID:3942492230
```

Fig. 14 An example user habit based on frequent stay points: user visits the "Baron cafe" on Sunday and Monday evenings

1. **MoveType = Metro ⇒ IsWorkingDay = true**
2. **DayZone = Evening Tag = public_transport 331 ⇒ IsWorkingDay = true**
3. **DayZone = Evening MoveType = Metro Tag = public_transport ⇒ IsWorkingDay = true**
4. **DayZone = Evening MoveType = Metro ⇒ IsWorkingDay = true**
5. **Tag = shop ⇒ IsWorkingDay = true**

...

...

Fig. 15 Sample of the extracted habits after the Apriori execution

$$\{DayZone, DayType, MoveType, Dest\_POI_{Type}\},$$

where $DayType$ can be weekday or weekend and the $Dest\,POI_{Category}$ can be the type of POI detected as frequent stay point (e.g. cafeteria), the extracted user habits are similar to those depicted in Fig. 15.

Based on the sample result of Fig. 15 we can assume for example, based on rules No. 2, 3 and 4 that the user tends to commute by metro in the workday evenings, while based on rule No. 5 user also tends to visit a shop on working days.

In order to demonstrate the methodology for the generation of alternative routes, we processed the GPS log files for a specific user, with the tools that we have developed so far. Using the frequent stay points detection tool, we located all the frequent stay points, including the user's home and workplace. Next, we applied the frequent trajectory detection algorithm, which revealed the daily commute trajectories for the specific user. Following the methodology that we introduced in Sect. 4, we select all the commute trajectories from the user Google Maps history and apply the partition and clustering methodology presented in this work. Using k-means clustering with k = 35, we resulted with 35 representative line segments, one for each of the created clusters. A simplified version of this graph (i.e. for k = 10 clusters) is depicted in Fig. 16.

By applying Yen's K-Shortest Path algorithm to the graph that summarizes all the commute trajectories, using $K = 5$, we get the five shortest commute paths for the specific user. The paths are ranked in increasing length order and are depicted in Fig. 17. The shortest length path is denoted as $1 \rightarrow 6 \rightarrow 8$, with $1 \rightarrow 5 \rightarrow 6 \rightarrow 8$ being the second shortest. The alternate paths can be used by a recommendation engine to propose alternative route choices to the user at any moment, using the current position as input.

7 Conclusions and Next Steps

Considering the implicit interaction among users who are sharing location information with other users, we can assume that these users form a type of social network with location-based information. In this work, we presented an application that processes user GPS logs to extract useful information that is enriched and abstracted, in order to extract rules and patterns that describe user's habits. These

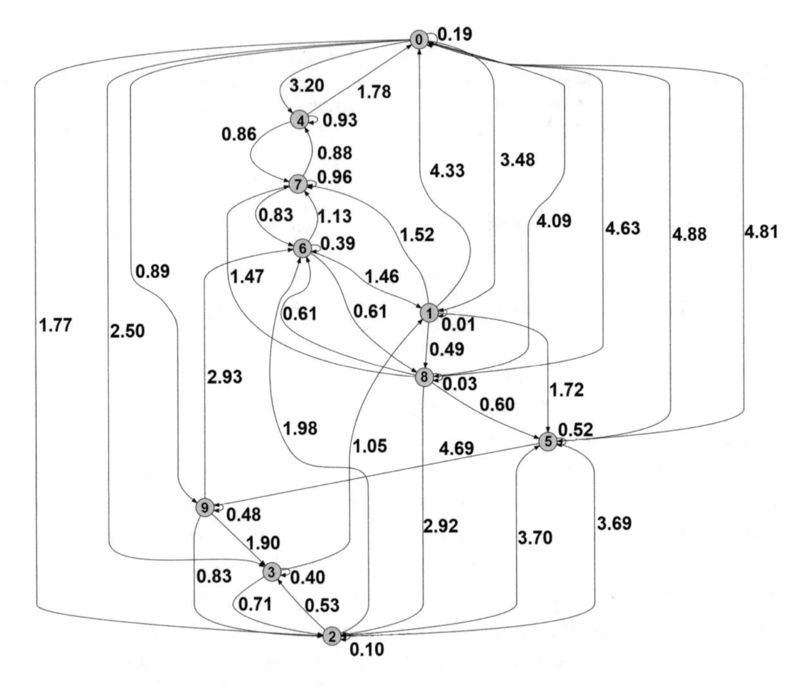

Fig. 16 The commute graph with all alternative routes

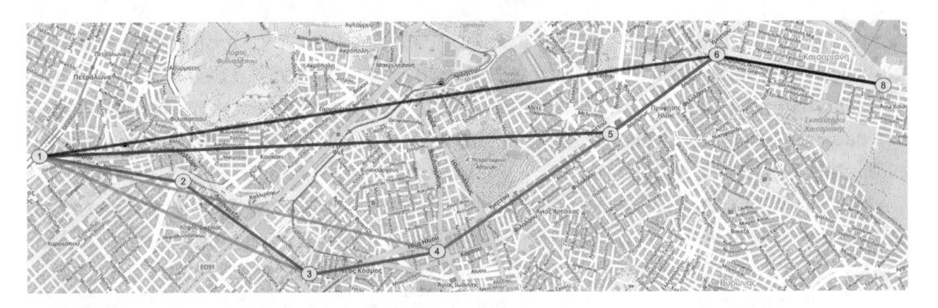

Fig. 17 The top-5 shortest alternative commute routes

behavioral patterns include frequently visited POIs, frequently used trajectories and associations among user preferences with day of week and/or timezone of activity. As far as the results show, we can use the GPS logs to identify interesting patterns for the user daily activity. This type of information could be exploited furthermore in many types of applications and fields of research, such as recommender systems for providing personalized recommendations.

Having that said, we consider this is a field of interest with lots of potential and it is in our intention to adapt our work so far in order to lead us to a recommender system that would deliver real-time and real-life recommendations based on user habits.

Moreover, the parameter selection of the algorithms has been made after experimentation on the specific dataset. A more thorough evaluation of different parameter settings is part of our next work on the field.

Acknowledgements This work has been developed in the frame of the MASTER project, which has received funding from the European Union's Horizon 2020 research and innovation programme under the Marie Skłodowska-Curie grant agreement No. 777695.

References

1. D. Bosomworth, *Mobile Marketing Statistics 2015* (Smart Insights (Marketing Intelligence) Ltd, Leeds, 2015)
2. P. Symeonidis, D. Ntempos, Y. Manolopoulos, Location-based social networks, in *Recommender Systems for Location-Based Social Networks* (Springer, Berlin, 2014), pp. 35–48
3. M. Eirinaki, J. Gao, I. Varlamis, K. Tserpes, Recommender systems for large-scale social networks: a review of challenges and solutions. Futur. Gener. Comput. Syst. **78**, 413–418 (2018). https://doi.org/10.1016/j.future.2017.09.015
4. W.-Y. Zhu, W.-C. Peng, L.-J. Chen, K. Zheng, X. Zhou, Modeling user mobility for location promotion in location-based social networks, in *Proceedings of the 21th ACM SIGKDD International Conference on Knowledge Discovery and Data Mining* (ACM, New York, 2015), pp. 1573–1582
5. Y. Yu, X. Chen, A survey of point-of-interest recommendation in location-based social networks, in *Workshops at the Twenty-Ninth AAAI Conference on Artificial Intelligence*, vol. 130 (2015)
6. P. Kouris, I. Varlamis, G. Alexandridis, A. Stafylopatis, A versatile package recommendation framework aiming at preference score maximization. Evol. Syst. 1–19 (2018). https://doi.org/10.1007/s12530-018-9231-2
7. Q. Yuan, G. Cong, Z. Ma, A. Sun, N.M. Thalmann, Time-aware point-of-interest recommendation, in *Proceedings of the 36th International ACM SIGIR Conference on Research and Development in Information Retrieval* (ACM, New York, 2013), pp. 363–372
8. C. Sardianos, I. Varlamis, G. Bouras, Extracting user habits from Google maps history logs, in *2018 IEEE/ACM International Conference on Advances in Social Networks Analysis and Mining (ASONAM)* (IEEE, Piscataway, 2018), pp. 690–697
9. Y. Zheng, Trajectory data mining: an overview. ACM Trans. Intell. Syst. Technol. **6**(3), 29 (2015)
10. Y. Zheng, L. Zhang, X. Xie, W.-Y. Ma, Mining interesting locations and travel sequences from GPS trajectories, in *Proceedings of the 18th International Conference on World Wide Web* (ACM, New York, 2009), pp. 791–800
11. Q. Li, Y. Zheng, X. Xie, Y. Chen, W. Liu, W.-Y. Ma, Mining user similarity based on location history, in *Proceedings of the 16th ACM SIGSPATIAL International Conference on Advances in Geographic Information Systems* (ACM, New York, 2008), p. 34
12. X. Cao, G. Cong, C.S. Jensen, Mining significant semantic locations from GPS data, Proc. VLDB Endowment **3**(1–2), 1009–1020 (2010)
13. H. Gao, J. Tang, X. Hu, H. Liu, Content-aware point of interest recommendation on location-based social networks, in *AAAI Conference on Artificial Intelligence* (2015), pp. 1721–1727
14. A.T. Palma, V. Bogorny, B. Kuijpers, L.O. Alvares, A clustering-based approach for discovering interesting places in trajectories, in *Proceedings of the 2008 ACM Symposium on Applied Computing* (ACM, New York, 2008), pp. 863–868
15. A.S. Furtado, L.L. Pilla, V. Bogorny, A branch and bound strategy for fast trajectory similarity measuring. Data Knowl. Eng. **115**, 16–31 (2018)

16. J.-G. Lee, J. Han, K.-Y. Whang, Trajectory clustering: a partition-and-group framework, in *Proceedings of the 2007 ACM SIGMOD International Conference on Management of Data* (ACM, New York, 2007), pp. 593–604
17. E. Oliveira, I.R. Brilhante, J.A.F. de Macedo, TrajectMe: planning sightseeing tours with hotel selection from trajectory data, in *Proceedings of the 2nd ACM SIGSPATIAL Workshop on Recommendations for Location-Based Services and Social Networks* (ACM, New York, 2018), p. 1
18. J. Bao, Y. Zheng, D. Wilkie, M. Mokbel, Recommendations in location-based social networks: a survey. Geoinformatica **19**(3), 525–565 (2015)
19. V.W. Zheng, Y. Zheng, X. Xie, Q. Yang, Collaborative location and activity recommendations with GPS history data, in *Proceedings of the 19th International Conference on World Wide Web* (ACM, New York, 2010), pp. 1029–1038
20. Y. Zheng, X. Xie, W.-Y. Ma, GeoLife: a collaborative social networking service among user, location and trajectory. IEEE Data Eng. Bull. **33**(2), 32–39 (2010)
21. E. Cho, S.A. Myers, J. Leskovec, Friendship and mobility: user movement in location-based social networks, in *Proceedings of the 17th ACM SIGKDD International Conference on Knowledge Discovery and Data Mining* (ACM, New York, 2011), pp. 1082–1090
22. J. He, X. Li, L. Liao, D. Song, W.K. Cheung, Inferring a personalized next point-of-interest recommendation model with latent behavior patterns, in *Thirtieth AAAI Conference on Artificial Intelligence* (2016)
23. D. Quercia, R. Schifanella, L.M. Aiello, The shortest path to happiness: recommending beautiful, quiet, and happy routes in the city, in *Proceedings of the 25th ACM Conference on Hypertext and Social Media* (ACM, New York, 2014), pp. 116–125
24. F. Sparacino, The museum wearable: real-time sensor-driven understanding of visitors' interests for personalized visually-augmented museum experiences, in *International Conference on Museums and the Web MW2002* (ERIC, 2002).
25. N. Bu, M. Okamoto, T. Tsuji, A hybrid motion classification approach for EMG-based human–robot interfaces using Bayesian and neural networks. IEEE Trans. Robot. **25**(3), 502–511 (2009)
26. I. Varlamis, Evolutionary data sampling for user movement classification, in *IEEE Congress on Evolutionary Computation (CEC)* (IEEE, Piscataway, 2015), pp. 730–737
27. S. Tragopoulou, I. Varlamis, M. Eirinaki, Classification of movement data concerning user's activity recognition via mobile phones, in *Proceedings of the 4th International Conference on Web Intelligence, Mining and Semantics (WIMS14)* (ACM, New York, 2014), p. 42
28. G. Butler, *Manage Your Mind: The Mental Fitness Guide* (Oxford University Press, New York, 2007)
29. N. Eagle, A.S. Pentland, Reality mining: sensing complex social systems. Pers. Ubiquit. Comput. **10**(4), 255–268 (2006)
30. D. Gubiani, M. Pavan, From trajectory modeling to social habits and behaviors analysis, in *Recent Trends in Social Systems: Quantitative Theories and Quantitative Models* (Springer, Berlin, 2017), pp. 371–385
31. L.O. Alvares, V. Bogorny, B. Kuijpers, J.A.F. de Macedo, B. Moelans, A. Vaisman, A model for enriching trajectories with semantic geographical information, in *Proceedings of the 15th Annual ACM International Symposium on Advances in Geographic Information Systems* (ACM, New York, 2007), p. 22
32. R. Krüger, D. Thom, T. Ertl, Semantic enrichment of movement behavior with foursquare–a visual analytics approach. IEEE Trans. Vis. Comput. Graph. **21**(8), 903–915 (2015)
33. Z. Yan, D. Chakraborty, C. Parent, S. Spaccapietra, K. Aberer, Semitri: a framework for semantic annotation of heterogeneous trajectories, in *Proceedings of the 14th International Conference on Extending Database Technology* (ACM, 2011, pp. 259–270)
34. V. Bogorny, C. Renso, A.R. de Aquino, F. de Lucca Siqueira, L.O. Alvares, Constant–a conceptual data model for semantic trajectories of moving objects. Trans. GIS **18**(1), 66–88 (2014)

35. C. Parent, S. Spaccapietra, C. Renso, G. Andrienko, N. Andrienko, V. Bogorny, M.L. Damiani, A. Gkoulalas-Divanis, J. Macedo, N. Pelekis et al., Semantic trajectories modeling and analysis. ACM Comput. Surv. **45**(4), 42 (2013)
36. C. Li, W.K. Cheung, J. Liu, J.K. Ng, Automatic extraction of behavioral patterns for elderly mobility and daily routine analysis. ACM Trans. Intell. Syst. Technol. **9**(5), 54 (2018)
37. M. Lv, L. Chen, Z. Xu, Y. Li, G. Chen, The discovery of personally semantic places based on trajectory data mining. Neurocomputing **173**, 1142–1153 (2016)
38. S. Kisilevich, F. Mansmann, D. Keim, P-DBSCAN: a density based clustering algorithm for exploration and analysis of attractive areas using collections of geo-tagged photos, in *Proceedings of the 1st International Conference and Exhibition on Computing for Geospatial Research & Application* (ACM, New York, 2010), p. 38
39. A.A. Shaw, N. Gopalan, Finding frequent trajectories by clustering and sequential pattern mining. J. Traffic Transp. Eng. (Engl. Ed.) **1**(6), 393–403 (2014)
40. R. Agarwal, R. Srikant et al., Fast algorithms for mining association rules, in *Proceedings of the 20th VLDB Conference* (1994), pp. 487–499
41. M. Etemad, A. Soares Júnior, S. Matwin, Predicting transportation modes of GPS trajectories using feature engineering and noise removal, in *31st Canadian Conference on Artificial Intelligence* (Springer, Berlin, 2018), pp. 259–264
42. J.Y. Yen, Finding the K shortest loopless paths in a network. Manag. Sci. **17**(11), 712–716 (1971)

Semi-Automatic Training Set Construction for Supervised Sentiment Analysis in Polarized Contexts

S. Martin-Gutierrez, J. C. Losada, and R. M. Benito

1 Introduction

Supervised machine learning classification is one of the most relevant and widely used techniques for sentiment analysis [1]. This methodology is based on training a classification algorithm with text samples that have been previously labeled as positive or negative (or neutral in some cases). The classification algorithm can then be fed with new text samples, which are assigned a positive or a negative sentiment.

Although this approach usually offers very good results [2, 3], it suffers from one burden: the training set of pre-labeled samples need, in most cases, to be built by hand. This implies that a group of human annotators need to work on a set of text samples to assign them a sentiment. Hence, the size of the training set will be limited by the time and work that the human annotators are willing to invest in this task. Moreover, it has been shown that the inter-annotator agreement is usually around 80% [4].

In order to address this issue, an alternative methodology was proposed in [5] to automatize the retrieval and labeling of text samples from Twitter. They proposed to build a training dataset by retrieving messages with *happy* emoticons and *sad* emoticons. They also downloaded *objective* tweets from user accounts belonging to newspapers and magazines.

Another approach was considered in [6] and [7]. These two complementary works employ a lexicon-based technique to expand a seed dataset in order to obtain an automatically labeled training set. In [6] Vania et al. propose to build a seed

S. Martin-Gutierrez (✉) · J. C. Losada · R. M. Benito
Grupo de Sistemas Complejos, Escuela Técnica Superior de Ingeniería Agronómica, Alimentaria y de Biosistemas, Universidad Politécnica de Madrid, Madrid, Spain
e-mail: samuel.martin@upm.es; juancarlos.losada@upm.es; rosamaria.benito@upm.es
http://www.gsc.upm.es/

M. Kaya et al. (eds.), *Putting Social Media and Networking Data in Practice for Education, Planning, Prediction and Recommendation*, Lecture Notes in Social Networks, https://doi.org/10.1007/978-3-030-33698-1_10

lexicon by translating an existing one developed for the English language and expand the translated seed lexicon by looking for different grammatical patterns related to sentiment and by analyzing co-occurrence of words. Starting from this lexicon, in [7] Wicaksono et al. build a seed labeled dataset by performing a sentiment assignment based in counting the number of positive and negative terms from the lexicon that appear in each document and assigning them the sentiment associated to the majority of the terms. Then, they iteratively train and test a Naive Bayes classifier in unlabeled text samples through the Expectation-Maximization framework and build the training set choosing documents with high probability of belonging either to the positive or the negative class according to the classifier.

In [8], we proposed a methodology to automatically build a training set in a political context. In this work, we generalize our methodology to any system that presents a high polarization among the actors. More specifically, in order to apply our methodology to a given system, the requirement is that there exist well defined confronted groups (or poles). Although this could seem like a very limiting condition, it should be noticed that in very recent times several social contexts have emerged that show behaviors compatible with this requirement and have attracted considerable attention [9–14]. For example, Brexit [15, 16], the Catalan independence issue in Spain [17–19] or the 2016 USA presidential election [20].

In order to assign a sentiment to a tweet we take into account who is the author of the tweet and check if a list of terms or keywords appear in it. Our working assumption is that, in a polarized context, tweets sent by users associated to a given pole will be positive if they mention terms (names, slogans, other users, etc.) associated to the same pole and negative if they mention terms of any other pole (we consider the possibility of a multipolar system). We have tested this methodology in two contexts that typically exhibit a high polarization among the users: an important football match and a political campaign.

The polarized (or even polarizing) character of a conversation about a relevant football match is a well known fact that has been studied in the context of Twitter [21, 22]. We have considered tweets extracted from a conversation around a football match held the 23 of December 2017 [23] between the two top-teams [24] of the Spanish football league: Real Madrid Club de Fútbol and Barcelona Fútbol Club. The poles of this system would be the supporters of each team.

With respect to the polarization associated to political contexts, there is abundant literature centered around the study of this phenomenon in social media. See for example [25], where they reference several works providing empirical evidence about how politically active web users tend to organize into insular, homogeneous communities segregated along partisan lines [26, 27]. Analogous results have been obtained in [28, 29]. We have built our dataset with tweets associated to the conversation around the Spanish general elections of 2015 and the repetition of the elections in 2016. In the context of political campaigning, the poles in which the users are organized correspond to the different political parties.

Notice that, although this work is focused on Twitter conversations in Spanish, there is no limitation to apply this technique to different languages. Hence, it may be useful for researchers that need to tackle under-resourced languages.

A training set built following this methodology may suffer from some bias: the proportion of positive to negative messages posted by the user sets defined as the poles may not be the same as the corresponding proportion for the bulk of users. Accordingly, we also discuss the situation in which the dataset used to train and test an algorithm and the dataset in which it is applied have different base rates of positive and negative samples. This discussion also holds importance beyond this particular application.

In order to test our methodology, we build training and test sets by combining reference datasets [30] composed of manually labeled samples with the automatically labeled dataset obtained by the application of our approach to the previously mentioned Twitter conversations. We then train and test a sentiment classifier proposed in [3].

2 Materials and Methods

We have summarized the methodology that we have adopted to build the training sets, from the retrieval of the data to the text processing, in Fig. 1. In this section we detail the whole process.

2.1 Description of the Datasets

We have worked with Twitter messages retrieved with the Twitter streaming API. This API allows to download *tweets* matching a set of keywords. We have chosen the following keywords to filter the messages:

- Keywords for the 2015 election: *20D, 20D2015, #EleccionesGenerales2015.*
- Keywords for the 2016 election: *26J, 26J2016, #EleccionesGenerales2016, #Elecciones26J.*
- Keywords for the Madrid-Barcelona football match: *madrid barsa, madrid barça, futbol madrid, fútbol madrid, futbol barcelona, fútbol barcelona, futbol barsa, fútbol barsa, fútbol barça, futbol barça, elclasico, elclásico, fcbarcelona, fcbarcelona_es, fcbarcelona_cat, realmadrid, clasico futbol, clásico futbol, clasico fútbol, clásico fútbol, clasico madrid, clásico madrid, clasico barsa, clasico barça, clásico barsa, clásico barça.*

In Fig. 2 the word clouds of the most frequent terms in the different datasets are shown.

In the case of the elections dataset we have downloaded tweets during a period that spans from 1 November 2015 to 19 December 2015 and from 1 April 2016 to 25 June 2016. Both data collection periods span just until the day before the election. The reason to avoid collecting tweets posted in the election day or at any time after that is that once the campaign is finished there is some exchange of cordial messages

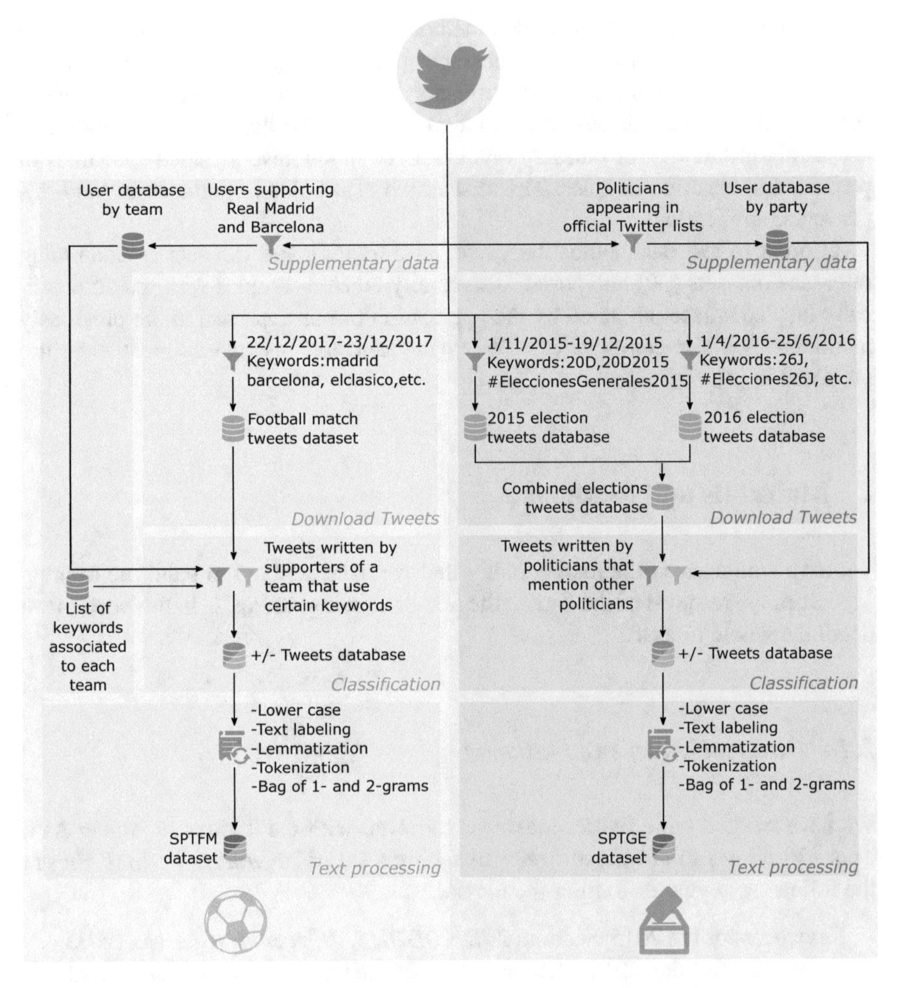

Fig. 1 Diagram that shows the process adopted to retrieve tweets associated to the conversations around the Spanish elections and the football match. The subsequent filtering to extract the training set for sentiment analysis and the text processing to standardize and tokenize it is also shown

between the politicians. Given the nature of our methodology, this would introduce noise in our automatically built training set.

In [31] we perform a thorough characterization of the general features of these conversations and, in particular, of the user behavior. In that work we show that there exist recurrent patterns of user behavior that are consistently manifested in both electoral periods.

We have also compiled lists of twitter accounts associated to the four main parties (PP, PSOE, Podemos and Cs). In order to build the set of political accounts, we have looked into the twitter lists (lists of accounts elaborated by the users) defined by relevant official accounts associated to each party. We have downloaded those lists

Fig. 2 Word clouds for the most used terms in each of the datasets: 2015 Spanish elections (left), 2016 Spanish elections (middle) and football match (right). The size is proportional to the frequency of each hashtag

that include politicians, political institutions or supporters of the party. The total number of retrieved political users that participated in the conversation was 5227 in 2015 and 5012 in 2016, with an average number of followers of 4044 (2015) and 4662 (2016).

The tweets corresponding to the conversation around the Madrid-Barcelona match have been downloaded from the 22 of December at 5 AM to the 23 of December at 13 PM (local time of Spain—UTC+1), which coincides with the time the match started. We stopped collecting tweets at the beginning of the match because the supporters of the losing team (in this case, Real Madrid), could potentially post negative tweets against their own team, rendering our working hypothesis useless.

We have compiled lists of accounts that follow either engaged users or official accounts of sports fan clubs with a strong association to Real Madrid or Barcelona. Specifically, we have retrieved 202,074 users that support Real Madrid and 183,874 users that support Barcelona by looking at the followers of these accounts:

- Accounts supporting Real Madrid: *aficion_Rmadrid, FondoSur_1980, GradaFansRMCF, RealMadrid_GO*
- Accounts supporting Barcelona: *madridblaugrana, Culedeleon, MDPilar6, PENYA_B_MADRID, UNIVERSO_1899*

We have also filtered out those users that followed at least one account associated to each of the teams, which amounted to 0.6% of the total.

In order to perform tests on our automatically built dataset, we have used two manually labeled datasets provided by the TASS organization (Spanish acronym for Workshop on Semantic Analysis at SEPLN-Spanish Society for Natural Language Processing).

The first one, which we have called TASS, is composed of 67,395 tweets (24,905 positive, 17,829 negative and 24,661 neutral) written in Spanish by 150 well-known personalities and celebrities of the world of politics, economy, communication, mass media and culture. It was built between November 2011 and March 2012.

The second, which is called STOMPOL (corpus of Spanish Tweets for Opinion Mining at aspect level about POLitics), contains 1220 political tweets (291 positive,

666 negative and 263 neutral) related to political aspects that appeared in the Spanish political campaign of regional and local elections that were held on 2015. These tweets were gathered from the 23 to the 24 of April 2015 [30].

The tweets contained in the TASS corpus also have labels according to the main topics of each message. We have exploited this feature to extract 534 positive and 152 negative tweets associated to football. We have called this selection of tweets TASSf.

2.2 *Automatized Construction of the Training Set*

In order to build a training set for sentiment classification of tweets, we normally would need to manually label an enormous quantity of messages or rely in the applicability of a third party corpus to our system. In order to avoid that, we have adopted some criteria to retrieve tweets whose sentiment can be easily inferred. These criteria are based on the following assumption: In a polarized context, users associated to a given pole will speak positively of their own pole and negatively of any other pole. In Fig. 3 we illustrate this rationale for the case of political polarization. The precise criteria for the assignment of sentiment must be adapted to the specific system under study.

Consequently, in the case of the elections, we have selected tweets sent by users appearing in twitter lists associated to several political parties and that include mentions to accounts belonging either to users of their own party or to users of

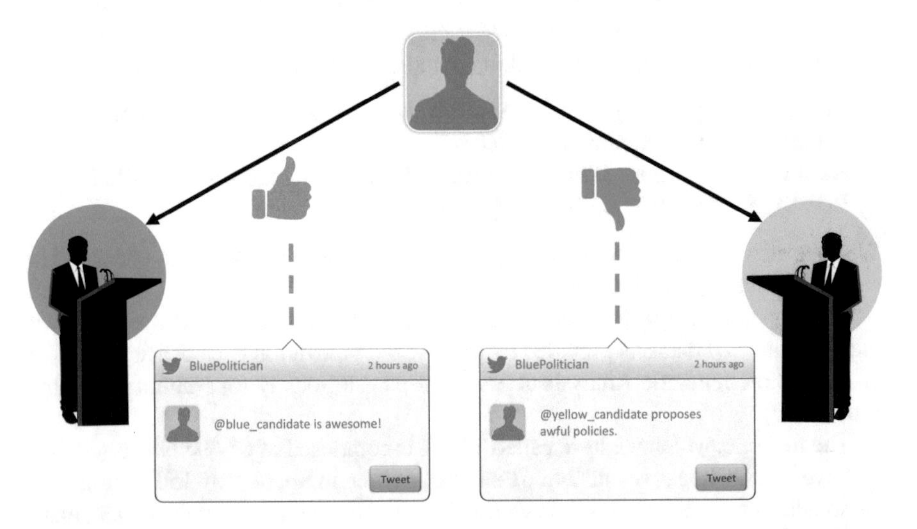

Fig. 3 Diagram that illustrates the idea that in a polarized context, members of a pole will post positive messages about other partners of their own pole and negative messages about members of different poles

different parties. Among these users we find politicians, organizations affine to specific parties, some sympathizers, etc. From these tweets we filter out retweets and keep only original tweets; otherwise we would end up with a degenerate dataset. We assign a sentiment to each message depending on who posts the message and who is mentioned in the tweet:

- Positive tweets: Those that include mentions only to accounts of the same party as the posting user.
- Negative tweets: Those that contain mentions to accounts of different parties and that do not contain mentions to accounts of the same party. They may include mentions to other accounts.

By adopting these criteria we obtained 18,902 positive tweets and 567 negative tweets. The scarcity of negative tweets is due to the lack of communication between different parties, which is a recurrent property of electoral contexts [31–33]. We have called this dataset Spanish Polarized Tweets of the General Elections (SPTGE).

With respect to the football match, we have selected tweets sent by users that follow accounts belonging either to engaged users or to sports fan clubs with a strong association to one of the teams. From these tweets, we have extracted those that contain keywords linked to one of the two teams. These keywords are the names and Twitter accounts of the players and the coaches that participated in the match and the names and some nicknames of the teams (*culés* for Barcelona and *merengues* for Real Madrid for example). Some of the keywords associated to Real Madrid are *Real Madrid, Ronaldo, Zidane, etc.* In the case of Barcelona, some examples of keywords are *Barcelona, Messi, Valverde, etc.* The criteria to assign a sentiment to a tweet are the following:

- Positive tweets: Those that include keywords associated only to the same team as the posting user.
- Negative tweets: Those that include keywords associated only to the rival team.

By adopting these criteria we obtained 1729 positive tweets and 273 negative tweets. We have called this dataset Spanish Polarized Tweets of a Football Match (SPTFM).

2.3 Sentiment Classification Algorithm

After the retrieval of the data, we have preprocessed and standardized the text in the following ways:

- The whole text has been converted to lower case.
- The metadata of the tweet have been used to identify mentions, media and urls and replace them with the *11mention11*, *11media11* and *11url11* labels respectively. If a url is present in the text of the tweet but not in the metadata, it is replaced also with *11url11*.
- The terms associated to any pole have also been substituted by *11mention11*. For example, *Rajoy* or *Pedro Sánchez* in the case of the elections dataset and *Zidane* or *Valverde* in the case of the football dataset.

- In order to standardize the most used cheers used by Real Madrid and Barcelona fans, *hala madrid* and *visca barça* respectively, we have replaced them by *11exclamation11*.
- The words of the text have been lemmatized with a dictionary for Spanish lemmatization [34].
- The text has been then tokenized by splitting it with the following regular expression:

```
[#¿\?¡!\.,;:\[\]\"\s“”-]+
```

  Any character that matches the regular expression is a splitting point.
- The split text has been then converted to a bag of unigrams and bigrams.

We have adopted the classification algorithm proposed in [3]. In that work, Wang et al. propose to train a multinomial naive Bayes (MNB) classifier and use its log-count ratios as feature vectors for a discriminative classifier such as a support vector machine (SVM) or a logistic regression (LR) classifier. Let us briefly describe their method.

Let $\boldsymbol{f}^{(i)} \in \mathbb{R}^{|\mathbb{V}|}$ be the feature count vector for training case i with label $y^{(i)} \in \{-1, 1\}$. V is the set of features (unigrams and bigrams in our case) and $\boldsymbol{f}_j^{(i)}$ represents the number of occurrences of feature V_j in training case i. Define the count vectors as $\boldsymbol{p} = \alpha + \sum_{i:y^{(i)}=1} \boldsymbol{f}^{(i)}$ and $\boldsymbol{q} = \alpha + \sum_{i:y^{(i)}=-1} \boldsymbol{f}^{(i)}$ for smoothing parameter α, which we have chosen to be $\alpha = 1$. The log-count ratio is:

$$\boldsymbol{r} = \log\left(\frac{\boldsymbol{p}/||\boldsymbol{p}||_1}{\boldsymbol{q}/||\boldsymbol{q}||_1}\right). \tag{1}$$

Let $\boldsymbol{x}^{(k)}$ be the feature vector that represents text sample (k). In order to build an MNB feature vector, we need first to obtain the feature count vector $\boldsymbol{f}^{(k)}$ and binarize it: $\hat{\boldsymbol{f}}^{(k)} = \mathbf{1}\{\boldsymbol{f}^{(k)} > 0\}$ where $\mathbf{1}$ is the indicator function. The MNB feature vector can then be computed as $\boldsymbol{x}^{(k)} = \boldsymbol{r} \circ \hat{\boldsymbol{f}}^{(k)}$, where $\circ$ is the element-wise product.

The obtained MNB feature vectors $\boldsymbol{x}^{(k)}$ are used then as input for a discriminative classifier. In our case, we have opted for a logistic regression (LR) classifier cross validating the C parameter and calibrated using sigmoid calibration [35, 36].

2.4 *Choosing a Classification Threshold for Unknown Base Rates*

A classification problem consists in assigning a label, out of a fixed set of classes, to a particular entity, which is characterized by a vector of features. Given the entity to be classified, machine learning algorithms usually work by assigning a score to each possible label, choosing the one with the highest score.

If the problem only has two possible classes (binary classification), one may proceed as follows: for each entity to be classified, subtract the score for class 2 from the score of class 1. If the result is above 0, the entity is assigned to class 1, otherwise it is assigned to class 2. This approach enables us to chose a different classification threshold. If we push it above 0, it will be more probable for samples to be classified as class 2 than as class 1, and vice versa.

A common approach to choose a classification threshold is to plot the precision-recall curve and decide which point of the curve (which corresponds to a specific threshold) is best suited for a given application. However, in some situations (as is our case), the training samples may have not been randomly sampled from the population. Hence, the base rates of members of each class may not be the same in the training and test sets as in the data where the algorithm is intended to be used (let us call it the application set). In this case, the precision is not very informative, as it depends heavily on the base rates. The recall however can be trusted, since its consistency is guaranteed as long as the new samples that are fed to the algorithm are *similar* to the samples of the training set. Let us express this formally.

Given a binary classification task with a number of positive samples $P = TP + FN$, a number of negative samples $N = TN + FP$ and an outcome of TN true negatives, TP true positives, FN false negatives and FP false positives, the precision p and the recall r are defined as follows:

$$p = \frac{TP}{TP + FP} \quad ; \quad r = \frac{TP}{TP + FN} = \frac{TP}{P} \,. \tag{2}$$

In Fig. 4 we present an intuitive visualization of a confusion matrix with its corresponding TN, TP, FN and FP quantities.

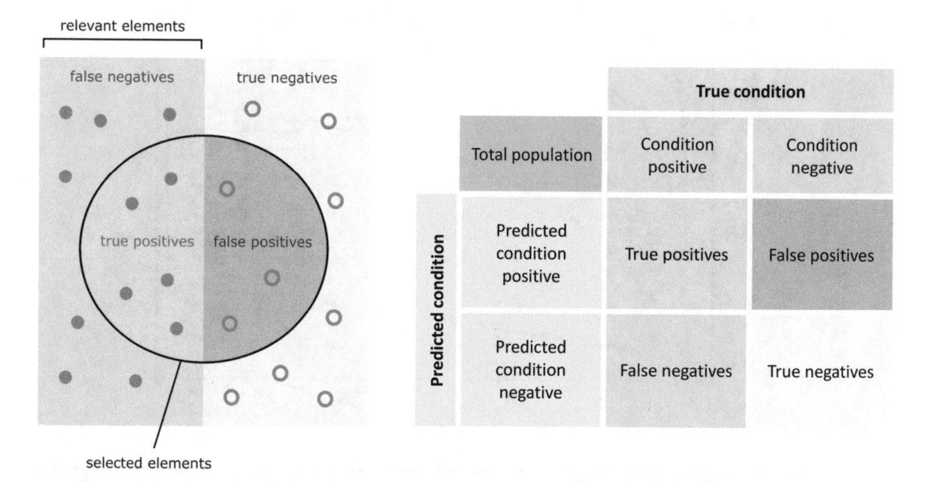

Fig. 4 Visualization of a confusion matrix in a binary classification task. This figure is a modified version of [37]

In (2) we see that the recall do not depend on the proportion of positives to negatives. It only informs about the proportion of correctly classified positives out of the total number of positives. Let us now express the precision in the following way:

$$p = \frac{rP}{N(1-s) + rP} = \frac{1}{\frac{N}{P}\frac{1-s}{r} + 1}, \tag{3}$$

where $s = \frac{TN}{N}$ is the specificity or true negative rate.

So p depends on the N/P ratio and on a function of both r and s. Since both quantities are positive, for a fixed N/P, if we wanted to improve our precision, we would need to minimize $\frac{1-s}{r}$; or, conversely, maximize $\frac{r}{1-s}$. This metric is the positive likelihood ratio (LR_+), and is insensitive to the positive to negative samples ratio. The reason is that, for a given classifier, if P (N) is increased or decreased, TP (TN) will increase or decrease in the same proportion, leaving r (s) constant. Consequently, we could use LR_+ as an alternative measure for the precision.

If we plot p as a function of LR_+ for different base rates; that is, different proportions of positive samples P,

$$p = \frac{1}{\frac{1-P}{P}\frac{1}{LR_+} + 1}, \tag{4}$$

we have the situation shown in Fig. 5. As we can see, when the proportion of positive samples P is too small, there is no way to achieve a high precision, even if the likelihood ratio is very good. This is a very inconvenient result: independently of

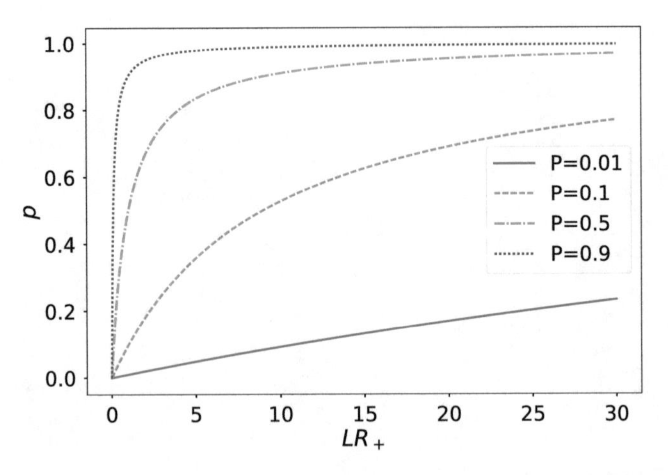

Fig. 5 Dependence of the precision p with respect to the proportion of positive samples P and the positive likelihood ratio LR_+. Although LR_+ is not bounded, the empirical results suggest that it is reasonable to expect that a classifier will not exceed a value of $LR_+ = 20$

how good our classifier is, if the entities of the class we aim to detect are scarce, we have no way to ensure that the identified samples actually belong to that given class.

Let us sum up the discussion above. When choosing a threshold for a classification task, we usually focus on balancing out the precision and recall of one of the classes. If the base rates of the dataset to be classified are unknown, the precision (computed with the test set) may not be informative. The likelihood ratio of the chosen class could be used instead.

A further step can be taken if we do not want to give up on using the precision: a sample of the application dataset can be extracted and the base rates can be computed by counting their occurrence by hand. In the case of tweets, this is not a very difficult task, since a lot of messages can be read in a relatively short time. Once the base rates are estimated, they can be plugged in (3) together with the recall and specificity calculated for the test set and obtain an approximated precision.

An alternative and potentially easier way would be to compute the corrected precision p^* for the application dataset directly from the original precision p computed for the test set and the N/P ratios of each dataset. To do this, we should take into account that LR_+ is the same for both datasets. Then, if we define $R^* = N^*/P^*$ as the negative to positive samples ratio for the application set and R as the corresponding ratio for the test set, the corrected precision is:

$$p^* = \frac{1}{\frac{R^*}{R}\left(\frac{1}{p} - 1\right) + 1}\,. \tag{5}$$

Then, a cross validation of the classifier can be performed such that a confidence interval Δp for the precision is obtained. Additionally, the uncertainty ΔR^* can be easily inferred by bootstrapping or by computation of binomial proportion confidence intervals [38]. With this information, an estimation of the confidence interval for p^* can be computed straight forward by error propagation:

$$\Delta p^* \approx \left(\frac{p^*}{p}\right)^2 \frac{R^*}{R}\Delta p + \frac{p^{*2}}{R}\left(\frac{1}{p} - 1\right)\Delta R\,. \tag{6}$$

Summing up, with Eq. (5) we have a way to estimate the corrected precision of a classifier trained and tested in a dataset whose base rates are biased with respect to the dataset where we intend to apply it. Equation (6) provides us with the means to determine the confidence interval of such corrected precision.

3 Results

3.1 Testing the Automatically Built Training Set

In order to test the proposed methodology to automatically build a training set for supervised sentiment analysis in a polarized context, we have compared our datasets with two manually labeled datasets provided by the TASS organization: a dataset of tweets covering several topics which we call TASS and a dataset containing political tweets called STOMPOL. We have also extracted those tweets with football-related content from the TASS corpus and built the TASSf dataset. These corpora have been described in Sect. 2.1.

When tweets are examined in a political context, sentiment analysis is usually used to verify if a text sample supports or criticizes a given politician or political party that is mentioned in the sample. The interpretation of a message classified as positive or negative is clear; however, a neutral message in a political context can not be considered to be in the middle point between a criticism and a praise. In these cases, neutral messages usually hold information about events like debates, meetings or rallies. Consequently, if posted by a common user, these kind of messages can be treated as positive, since that user is spreading information that can potentially benefit the political party mentioned in the message. Hence, for the elections dataset, we combine neutral and positive messages in the same class. In the case of the football dataset, we have kept only positive and negative tweets.

To perform the comparison between the different datasets, we have carried out a kind of cross validation of our datasets (SPTGE and SPTFM) by using the complementary datasets mentioned above. In order to do that, we have iterated over the following procedure:

1. Choose a given dataset (SPTGE, STOMPOL or TASS for the political context and SPTFM or TASSf for the football match) or a combination of several datasets.
2. Split the dataset in train/test applying stratified K-fold with $K = 3$: use 1/3 of the samples for test and maintain the proportion of negative and positive samples in the train and test sets.
3. Train the classifier described in Sect. 2.3 and test it in the corresponding sets, computing the area under the receiver-operating characteristic curve (ROC) [39] to check the quality of the classification.
4. If a combination of datasets has been chosen, compute the area under the ROC individually for the test samples corresponding to each individual dataset.
5. Test the classifier on the remaining individual datasets which have not been used to train the classifier.
6. Repeat steps 2–5 for each of the three possible train/test splits and obtain average values and standard deviations for the area under the ROC.
7. Repeat steps 1–6 for every possible choice of datasets (single and combination).

The resulting average values of areas under the ROC of these tests for the political context are presented in Table 1. The uncertainty corresponds to one standard deviation (1σ).

As one would expect, when we train only in one corpus, the best test result corresponds to the samples obtained from the same dataset. It is so because, even if they have not been seen by the algorithm, they are more similar to the training samples.

Another relevant characteristic of the individual datasets that boosts their ROC is their size. The best classification is the one obtained with TASS (67,395 tweets), followed by SPTGE (19,469 tweets) and STOMPOL (1220).

When we train the classifier in a given dataset and test it in a different one, the obtained area under the ROC is lower. Nevertheless, we should take into account that sentiment prediction models are known to be sensitive to the domain [30]. This is relevant because, on the one hand, every dataset was compiled in a different context. On the other hand, STOMPOL and SPTGE contain only political tweets, while TASS is general (and includes political tweets). With these facts in mind, the results of the tests are satisfactory.

One of the most relevant testings that we have performed to assess the correctness of our method is the combination of several datasets in a single training set followed by a testing performed in each of the individual sets of such combination. This analysis is based upon the idea that, if the SPTGE dataset is meaningful, when we combine it with other similar datasets, the classification should yield comparable results as the classification carried out with the individual datasets.

As we can appreciate in the bottom four rows of Table 1, the quality of the classification when the algorithm is trained in a combination of datasets is comparable to the results obtained for individual training. Moreover, the quality of the classification slightly improves for SPTGE when it is combined with STOMPOL.

When the classifier is trained in the SPTGE dataset (or a combination of SPTGE and STOMPOL) and tested in the TASS dataset, the area under the ROC takes values as low as a random classifier. However, as we discussed earlier, sentiment classification is known to be domain dependent and, while dataset TASS contains samples of several topics, SPTGE and STOMPOL contain only political tweets. Consequently, since the classifier has not seen samples similar to many tweets that can be found in TASS, the quality of the classification can be expected to be low.

The higher quality of a classifier trained in the STOMPOL dataset and tested in TASS could imply that the text samples present in SPTGE are slightly more specific than those of TASS (which is obvious) and STOMPOL. To add further evidence, notice that the SPTGE dataset is also pretty well classified by an algorithm trained in TASS or STOMPOL.

Although a combination of SPTGE + TASS seems to achieve the best general results when compared with other combination of datasets, notice that the obtained area under the ROC is the same as for the TASS dataset trained individually. In order to perform a correct comparison, we should also look at the area under the ROC for the test samples of the individual datasets, which are high for every dataset when

Table 1 Average area under the receiver-operating characteristic curve (ROC) for a stratified three-fold cross validation for the electoral context datasets

Train ↓/Test →	SPTGE	STOMPOL	TASS	Combination
SPTGE	0.823 ± 0.013	0.640 ± 0.008	0.50 ± 0.03	–
STOMPOL	0.631 ± 0.010	0.77 ± 0.02	0.650 ± 0.004	–
TASS	0.6534 ± 0.0015	0.673 ± 0.012	0.908 ± 0.002	–
STOMPOL + TASS	0.665 ± 0.004	0.726 ± 0.013	0.907 ± 0.003	0.903 ± 0.004
SPTGE + TASS	0.81 ± 0.02	0.680 ± 0.006	0.904 ± 0.002	0.908 ± 0.002
SPTGE + STOMPOL	0.847 ± 0.008	0.73 ± 0.02	0.526 ± 0.008	0.899 ± 0.003
SPTGE + STOMPOL + TASS	0.82 ± 0.02	0.74 ± 0.03	0.9030 ± 0.0008	0.9046 ± 0.0008

The model is trained in a single dataset or a combination of them. Then it is tested in unseen samples of every corpus (including unseen samples from the corpus used as training set). The Combination column corresponds to the area under the ROC for a test set composed of samples of the combination of corpora used as training set (in the case of a single corpus it coincides with the value obtained for the corresponding test set of the same name). The uncertainty corresponds to 1σ

Table 2 Average area under the receiver-operating characteristic curve (ROC) for a stratified three-fold cross validation for the football context datasets

Train/Test	SPTFM	TASSf	Combination
SPTFM	0.756 ± 0.013	0.64 ± 0.03	–
TASSf	0.628 ± 0.010	0.82 ± 0.02	–
SPTFM + TASSf	0.760 ± 0.004	0.80 ± 0.05	0.77 ± 0.02

The model is trained in a single dataset or a combination of them. Then it is tested in unseen samples of every corpus (including unseen samples from the corpus used as training set). The Combination column corresponds to the area under the ROC for a test set composed of samples of the combination of corpora used as training set (in the case of a single corpus it coincides with the value obtained for the corresponding test set of the same name). The uncertainty corresponds to 1σ

the classifier is trained in a combination of the three corpora. Consequently, as we would expect, the algorithm generalizes best when trained with more samples and in more diverse contexts.

With respect to the football dataset, the average values of areas under the ROC of the tests are shown in Table 2. The uncertainty corresponds to one standard deviation (1σ).

The results that we have obtained with these datasets are analogous to those obtained with the political context. When we use samples from one corpus to train our classifier, the best result is obtained for unseen samples of the same corpus. In this case, however, TASSf, although smaller, yields a slightly better classification quality than SPTFM. The quality of the cross-classifications are comparable and satisfactory.

If we look at the combination training set of SPTFM and TASSf, we see that the results that we obtain when we test in the individual datasets are similar to those achieved when training only in samples from each of them individually. Moreover, consistently with the results obtained for the political context, the area under the ROC for test samples taken from the combination is also high.

As a general result, we can see that our datasets (SPTGE and SPTFM) offer comparable classification quality when they are compared with manually labeled datasets. This indicates that their samples and labeling are meaningful.

3.2 *Choosing a Classification Threshold for Biased Base Rates*

As we discussed in Sect. 2.4, we will find some issues if we train and test a classifier in a dataset whose samples are representative of the set we want to classify, but which do not have the same base rates. In these cases, the classifier could suffer from a bias and tend to classify samples from one class as if they belong to another one.

In principle, we could fix this by plotting the precision-recall curve for the class we are interested in and choosing the classification threshold that better suits our needs. However, when the base rates of the test set are not the same as those of the application set, the precision yields biased results. Fortunately, if we are able to estimate the base rates of the application set, we can compute an approximation of the true precision using (5).

We have tested this approach by combining all the considered datasets for each of the two studied systems; that is, we have taken the combination of SPTGE + TASS + STOMPOL (let us call it C1) on one hand and SPTFM + TASSf (C2) on the other. Then, we have split the sets in a training + test set and reserved a certain quantity of samples for a validation set.

In the case of the C1 dataset, the training + test sets are composed of 59,491 positive samples and 9531 negative samples and the validation set is composed of 9531 positive samples and 9531 negative samples. Hence, the training + test set has a positive to negative ratio of $P/N = 6.24$ and the validation set has a ratio of $P/N = 1$.

With respect to the C2 dataset, the training+test sets are composed of 2051 positive samples and 213 negative samples and the validation set is composed of 212 positive samples and 212 negative samples. Hence, the training+test set has a positive to negative ratio of $P/N = 9.67$ and the validation set has a ratio of $P/N = 1$.

We have then split the training + test set using a stratified three-fold, such that the test set contains $1/3$ of the samples and maintains the P/N ratio. Next, we have trained the classifier on the training set and tested it both in the test set and in the validation set. This step has been performed for the three possible partitions of the train + test dataset. Finally, we have computed the precision-recall curve for the test set and for the validation set. Additionally, we have computed the estimation of the precision-recall curve for the validation set calculating the precision by applying (5) to the results obtained with the test set and taking into account the P/N ratios of the validation and test sets. The described process has been performed 100 times, obtaining a total of 300 realizations when each of the three-fold splits are taken into account. Each time, the train + test/validation and the subsequent train/test splits have been randomized.

In Fig. 6 we have represented the average of the precision-recall curves for the 300 realizations in the C1 = TASS + STOMPOL + SPTGE dataset (left panel) and the C2 = SPTFM + TASSf dataset (right panel). The shadowed areas around the curves correspond to an uncertainty of 1σ. As we can appreciate in Fig. 6, the estimation of the precision once the base rates are taken into account is very reliable. Consequently, we could train and test a classifier in a dataset with biased base rates with respect to the application set and still obtain a good estimation of the precision-recall curve. This final step enables us to choose a classification threshold suited for our particular needs.

As can be noticed in Fig. 6, the precision for the test set is higher than for the validation set, since the proportion P/N is higher for the former. However, if we used the precision of the test set to choose a classification threshold, we would probably make a poor choice.

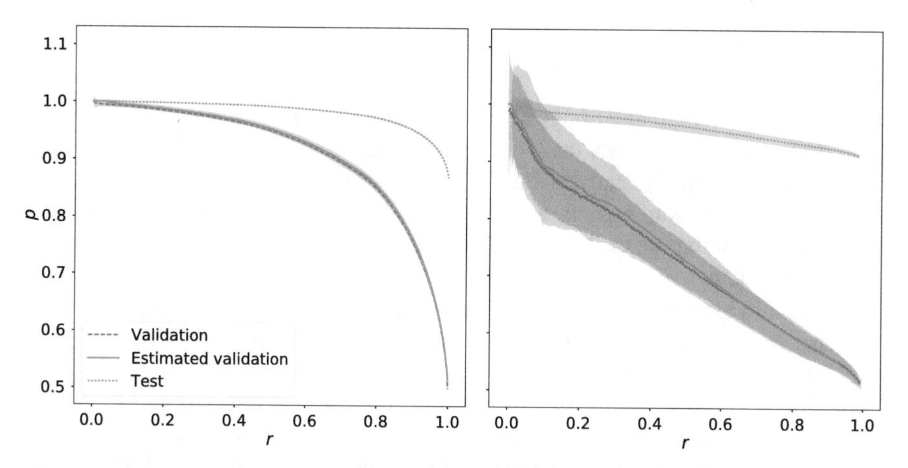

Fig. 6 Precision-recall curves for the test sets and the validation sets of the dataset corresponding to TASS + STOMPOL + SPTGE (left panel) and SPTFM + TASSf (right panel). The estimated precision curve applying (5) to the results for the test set and taking into account the base rates is also shown. The curves have been computed averaging 300 realizations. The shadowed areas correspond to 1σ

For example, in the case of the C1 dataset, if we looked for a compromise between high recall r and high precision p, we could choose the threshold associated to $r \approx 0.94$ and $p \approx 0.94$. But then, the real precision of the classifier would be $p^* \approx 0.7$.

In the case of the C2 dataset, the situation is even more dangerous due to the higher P/N ratio in the train+test set. In this scenario, a threshold with a recall of $r \approx 0.80$ and a $p \approx 0.93$ in the test set would yield a corrected precision of $p^* \approx 0.6$ in the validation set, which may be an issue depending on the use case of the classifier. Notice, however, that the reverse situation is also possible: we may get a better precision if the P/N ratio of the application set is higher than in the train+test set.

It can also be appreciated that the football dataset exhibits a precision-recall curve with larger deviations around the mean and a faster decay. The first characteristic is due to the size of the dataset, which is composed only of a few hundred samples, with respect to the several thousands of the political context dataset. This causes large fluctuations in the quality of the classification depending on which samples are used to train and which to test. With respect to the second feature, it can also partially be attributed to the size of the dataset (we have already discussed the importance of a large dataset in order to achieve a high classification quality), but another decisive factor is the P/N ratio of the train+test set, which is larger than the one obtained in the political context. A higher P/N ratio in the train + test set implies, for a fixed precision computed with the test set, a lower value of the corrected precision.

4 Conclusions

We have proposed a methodology to automatically build a training set for supervised sentiment analysis with data retrieved from Twitter in polarized contexts. Given the relevance of sentiment analysis for the research of online user behavior, this has the potential for becoming a powerful tool. The reason is that it enables researchers to build big training datasets with a fraction of the effort needed for labeling text samples by hand.

In order to test the applicability of the datasets built this way, we have performed a series of tests using a classification algorithm based on a combination of a generative model (multinomial naive Bayes) and a discriminative model (logistic regression). For performing these tests, two reference datasets provided by the Spanish Society for Natural Language Processing (SEPLN) were employed.

The tests consisted in training the classifier in one (or a combination) of the three datasets. Then it was tested both in unseen samples extracted from the same dataset as the one used for training and in the remaining datasets. Our findings show that the labeled samples obtained by our method are meaningful and can be trusted to be used as training set.

Complementarily, we have discussed how a classifier trained and tested in a dataset extracted from a given population but with biased base rates with respect to that population can be handled to compute an unbiased value of the precision. This discussion has enabled us to compute a corrected precision-recall curve that can be used to choose a suitable classification threshold. The developed formalism holds importance beyond this particular application.

The proposed methodology can be applied to any situation in which partisan behaviors emerge. Besides the examples provided in this work (political campaigns and sports), other scenarios in which this methodology may find application are those when different companies compete to promote similar products (Samsung vs Apple, Intel vs AMD, McDonalds vs Burger King, etc.).

Summing up, the proposed methodology to automatically build a training set for supervised sentiment analysis in polarized contexts has been proven to produce meaningful and well-labeled text samples. This technique complemented with the computation of an unbiased precision-recall curve has the potential to enable researchers to generate large and reliable training sets with a fraction of the effort needed to label text samples by hand.

Acknowledgements We would like to thank the TASS organization for allowing us to use their sentiment analysis corpus. This work has been supported by the Spanish Ministry of Economy and Competitiveness (MINECO) under Contract No. MTM2015-63914-P and by the Spanish Ministry of Science, Innovation and Universities (MICIU) Contract No. PGC2018-093854-B-100.

References

1. W. Medhat, A. Hassan, H. Korashy, Sentiment analysis algorithms and applications: a survey. Ain Shams Eng. J. **5**(4), 1093–1113 (2014) [Online]. http://www.sciencedirect.com/science/article/pii/S2090447914000550
2. P. Yang, Y. Chen, A survey on sentiment analysis by using machine learning methods, in *2017 IEEE 2nd Information, Technology, Networking, Electronic and Automation Control Conference (ITNEC)* (IEEE, Piscataway, 2017), pp. 117–121
3. S. Wang, C.D. Manning, Baselines and bigrams: simple, good sentiment and topic classification, in *Proceedings of the 50th Annual Meeting of the Association for Computational Linguistics: Short Papers – Volume 2*, ser. ACL '12 (Association for Computational Linguistics, Stroudsburg, 2012), pp. 90–94 [Online]. http://dl.acm.org/citation.cfm?id=2390665.2390688
4. V. Bobicev, M. Sokolova, Inter-annotator agreement in sentiment analysis: machine learning perspective, in *Proceedings of the International Conference Recent Advances in Natural Language Processing, RANLP 2017* (2017), pp. 97–102
5. A. Pak, P. Paroubek, Twitter as a corpus for sentiment analysis and opinion mining, in *Proceedings of the Seventh International Conference on Language Resources and Evaluation (LREC'10)*, vol. 10 (2010)
6. C. Vania, M. Ibrahim, M. Adriani, Sentiment lexicon generation for an under-resourced language. Int. J. Comput. Linguistics Appl. **5**(1), 59–72 (2014)
7. A.F. Wicaksono, C. Vania, B. Distiawan, M. Adriani, Automatically building a corpus for sentiment analysis on indonesian tweets, in *Proceedings of the 28th Pacific Asia Conference on Language, Information and Computing* (2014), pp. 185–194
8. S. Martin-Gutierrez, J.C. Losada, R.M. Benito, Semi-automatic training set construction for supervised sentiment analysis in political contexts, in *2018 IEEE/ACM International Conference on Advances in Social Networks Analysis and Mining (ASONAM)* (IEEE, Piscataway, 2018), pp. 715–720
9. G. Olivares, J.P. Cárdenas, J.C. Losada, J. Borondo, Opinion polarization during a dichotomous electoral process. Complexity **2019**, 9 (2019)
10. M. Hürlimann, B. Davis, K. Cortis, A. Freitas, S. Handschuh, S. Fernández, A Twitter sentiment gold standard for the brexit referendum, in *SEMANTiCS 2016 Proceedings of the 12th International Conference on Semantic Systems* (2016), pp. 193–196
11. M.T. Bastos, D. Mercea, The brexit botnet and user-generated hyperpartisan news. Soc. Sci. Comput. Rev. **37**(1), 38–54 (2019)
12. M.E. Del Valle, R.B. Bravo, Echo chambers in parliamentary Twitter networks: the Catalan case. Int. J. Commun. **12**, 21 (2018)
13. F. Guerrero-Solé, Community detection in political discussions on Twitter: an application of the retweet overlap network method to the Catalan process toward independence. Soc. Sci. Comput. Rev. **35**(2), 244–261 (2017)
14. U. Yaqub, S.A. Chun, V. Atluri, J. Vaidya, Analysis of political discourse on Twitter in the context of the 2016 US presidential elections. Gov. Inf. Q. **34**(4), 613–626 (2017)
15. S.B. Hobolt, T. Leeper, J. Tilley, *Divided by the Vote: Affective Polarization in the Wake of Brexit* (American Political Science Association, Boston, 2018)
16. M. Del Vicario, F. Zollo, G. Caldarelli, A. Scala, W. Quattrociocchi, Mapping social dynamics on facebook: the brexit debate. Soc. Net. **50**, 6–16 (2017)
17. D. Martí, D. Cetrà, The 2015 Catalan election: a de facto referendum on independence? Reg. Fed. Stud. **26**(1), 107–119 (2016)
18. A. Barrio, J. Rodríguez-Teruel, Reducing the gap between leaders and voters? elite polarization, outbidding competition, and the rise of secessionism in catalonia. Ethn. Racial Stud. **40**(10), 1776–1794 (2017)
19. I. Serrano, Just a matter of identity? Support for independence in Catalonia. Reg. Fed. Stud. **23**(5), 523–545 (2013)

20. P. Grover, A.K. Kar, Y.K. Dwivedi, M. Janssen, Polarization and acculturation in US election 2016 outcomes–can Twitter analytics predict changes in voting preferences. Technol. Forecast. Soc. Chang. **145**(C), pp. 438–460 (2018)
21. D.F. Pacheco, F. Lima-neto, L.G. Moyano, R. Menezes, Football conversations: what Twitter reveals about the 2014 world cup, in *Brazilian Workshop on Social Network Analysis and Mining (CSBC 2015-BraSNAM), Recife* (2015)
22. Z. Liu, I. Weber, Predicting ideological friends and foes in Twitter conflicts, in *Proceedings of the 23rd International Conference on World Wide Web* (ACM, New York, 2014), pp. 575–576
23. "Partido real madrid - fc barcelona en directo," online, accessed 10-December-2018 [Online]. https://www.laliga.es/directo/temporada-2017-2018/laliga-santander/17/real-madrid_barcelona
24. Wikipedia, "Anexo: Clubes españoles de fútbol ganadores de competiciones nacionales e internacionales — Wikipedia, the free encyclopedia," (2018), online, accessed 10-December-2018. [Online]. https://es.wikipedia.org/wiki/Anexo:Clubes_espa%C3%B1oles_de_f%C3%BAtbol_ganadores_de_competiciones_nacionales_e_internacionales
25. M. Conover, J. Ratkiewicz, M.R. Francisco, B. Gonçalves, F. Menczer, A. Flammini, Political polarization on Twitter. *Fifth International AAAI Conference on Weblogs and Social Media*, vol. 133, pp. 89–96 (2011)
26. L.A. Adamic, N. Glance, The political blogosphere and the 2004 US election: divided they blog, in *Proceedings of the 3rd International Workshop on Link Discovery* (ACM, New York, 2005), pp. 36–43
27. E. Hargittai, J. Gallo, M. Kane, Cross-ideological discussions among conservative and liberal bloggers. Public Choice **134**(1–2), 67–86 (2008)
28. M.D. Conover, B. Gonçalves, A. Flammini, F. Menczer, Partisan asymmetries in online political activity. EPJ Data Sci. **1**(1), 6 (2012)
29. A. Morales, J. Borondo, J.C. Losada, R.M. Benito, Measuring political polarization: Twitter shows the two sides of venezuela. Chaos Interdisciplinary J. Nonlinear Sci. **25**(3), 033114 (2015)
30. E.M. Cámara, M.A.G. Cumbreras, J.V. Román, J.G. Morera, Tass 2015 – the evolution of the spanish opinion mining systems. Procesamiento del Lenguaje Natural **56**, 33–40 (2016) [Online]. http://journal.sepln.org/sepln/ojs/ojs/index.php/pln/article/view/5284
31. S. Martin-Gutierrez, J.C. Losada, R.M. Benito, Recurrent patterns of user behavior in different electoral campaigns: a Twitter analysis of the Spanish general elections of 2015 and 2016. Complexity **2018**, 2413481 (2018) [Online]. https://doi.org/10.1155/2018/2413481
32. J. Borondo, A.J. Morales, J.C. Losada, R.M. Benito, Characterizing and modeling an electoral campaign in the context of Twitter: 2011 Spanish presidential election as a case study. Chaos Interdisciplinary J. Nonlinear Sci. **22**(2), 023138 (2012) [Online]. http://aip.scitation.org/doi/abs/10.1063/1.4729139
33. J. Borondo, A. Morales, R. Benito, J. Losada, Multiple leaders on a multilayer social media. Chaos, Solitons Fractals **72**, 90–98 (2015)
34. J. Atserias, B. Casas, E. Comelles, M. González, L. Padró, and M. Padró, Freeling 1.3: syntactic and semantic services in an open-source NLP library, in *Proceedings of the Fifth International Conference on Language Resources and Evaluation (LREC 2006)* (ELRA, Genoa, 2006)
35. F. Pedregosa, G. Varoquaux, A. Gramfort, V. Michel, B. Thirion, O. Grisel, M. Blondel, P. Prettenhofer, R. Weiss, V. Dubourg, J. Vanderplas, A. Passos, D. Cournapeau, M. Brucher, M. Perrot, E. Duchesnay, Scikit-learn: machine learning in Python. J. Mach. Learn. Res. **12**, 2825–2830 (2011)
36. A. Niculescu-Mizil, R. Caruana, Predicting good probabilities with supervised learning, in *Proceedings of the 22nd International Conference on Machine Learning*, ser. ICML '05 (ACM, New York, 2005), pp. 625–632. [Online]. http://doi.acm.org/10.1145/1102351.1102430
37. Walber, "File:precisionrecall.svg," Last accessed 11-July-2019. [Online]. https://en.wikipedia.org/wiki/File:Precisionrecall.svg

38. S. Wallis, Binomial confidence intervals and contingency tests: mathematical fundamentals and the evaluation of alternative methods. J. Quan. Linguist. **20**(3), 178–208 (2013)
39. K.A. Spackman, Signal detection theory: valuable tools for evaluating inductive learning, in *Proceedings of the Sixth International Workshop on Machine Learning* (Elsevier, Amsterdam, 1989), pp. 160–163

Detecting Clickbait on Online News Sites

Ayşe Geçkil, Ahmet Anıl Müngen, Esra Gündoğan, and Mehmet Kaya

1 Introduction

With the rapid development of technology and the internet, almost every sector has been moved to the internet and reached a broad audience in the digital world. Each official or private organization now has a branch on the web or an account on social media, as well as brand-new businesses working on the internet. Invoices are paid online, or most purchases are made online. Famous brands have opened internet branches of their stores and started to offer campaigns and opportunities specific to internet stores. Without leaving their homes, users get what they need at a much more affordable price, and products are brought to the door. For this reason, most of the internet sales are even ahead of store sales.

As in many other fields, we can say that journalism will become completely virtual over time. According to the Newspaper Spread Timeline [1], published by the Future Search Network, based on forecasts of printed newspapers, the last time the newspapers were published is as follows: in the United States in 2017, in England in 2017, and in Hong Kong in 2022, in France in 2029, in Turkey and Russia in 2036. So when the last newspaper in the world is published, the calendar will show the year 2040. Therefore, taking these estimates into account, all newspapers will be moved to the fully digital world in 2040.

Today, companies trying to make money from digital media and social networks that use it for their digital services. The revenues from digital publishing should be higher than the revenue from traditional services like printing media sales to continuing this digitalization. The Internet has its own rules and ways like it is not appropriate requesting a subscription fee for websites or magazines usually

A. Geçkil · A. A. Müngen · E. Gündoğan · M. Kaya (✉)
Department of Computer Engineering, Fırat University, Elazığ, Turkey
e-mail: egundogan@firat.edu.tr; kaya@firat.edu.tr

M. Kaya et al. (eds.), *Putting Social Media and Networking Data in Practice for Education, Planning, Prediction and Recommendation*, Lecture Notes in Social Networks, https://doi.org/10.1007/978-3-030-33698-1_11

for reading the news. Users probably prefer free of charge services instead of paid digital subscription. Both Digital News Media or Social Networks like YouTube media, it is required to get advertisements from companies or advertisers to develop content for digital platforms. Usually, companies pay by proportional to visitor traffic on web pages or social media platforms. In other words, the more visitor number in a certain period provide showing more ads, so this will also increase the ads revenue. The number of unique visitors will be announced on a daily basis for the advertisements companies. Websites, which are advertised in revenue items, are trying to increase their unique visitors focus by finding the fees they earn from the advertising cake they earn.

Newspapers, magazines, or news sites are receiving social media support to attract visitors to their pages and keep them on the page as long as possible. If we look at the pace of social media use, this support is too much to ignore. According to We Are Social's January 2017 data, 37% of the world's population (2 billion 789 million people) is an active social media user. Compared to 2015, 482 million users has increased. Also, the number of mobile social media users has reached 30% by raising 581 million. The number of social media and mobile social media users increased more than twice the number reported last year.

If web sites want to attract visitors using this power of social media, they have to choose the right news headlines and prepare them so that the visitor will be attracted and clicked on the link. This is because it is necessary to find the headline of the news to enter the site by sharing the news site that a user is a social media follower. Particularly when shared character limitations are considered, there is usually only the headlines of the news. Other news headlines on the rest of the page should be equally appealing to keep visitors on the site longer because the part of the news that appears on the page by the user is the title.

To ensure that the page is visited, it is not morally right to cast titles that are independent of the content, exaggerated or even untrue. Today, many newspapers or news sites frequently refer to this method in social media accounts as well as news headlines on official sites. Actions such as choosing exclamation expressions distant from the content of the news, related to a very small part of the society contrary to claim a large part to attract attention and display the news more interesting than it normally is, intentionally interrupting the headlines in the middle and using intensive pronouns instead of nouns focus on the style rather than the content, and decrease quality of the news. However, the widespread of this act was not averted. This act was named conceptually as Clickbait and took its place in dictionaries and literature. Simply, the act of using deceptive headlines to create a feeling of the relation between the content and the reader with the aim of clicking on the news link is called Clickbait.

There are multiple ways to create clickbait headlines. These methods are given in Table 1 with examples.

The Clickbait act is performed to cause the readers to think that they are related to the subject and ensure that they click on the link to the news. The most negative aspect of the Clickbait news is that some are not even reliable. It has been compiled from anywhere and added to the content due to its attractiveness rather than its

Table 1 Clickbait methods and examples

Clickbait method	Example
Using exaggerated exclamation expressions	PTT warned: Do not believe!
By directing you to a long gallery	10 reasons to consume palm!
Using pronouns instead of direct names	WhatsApp won't work on *these* phones!
Not completing the sentences	The winner of the Premier League...
By exaggerating the news mass that it is related to	Announced the decision of the world!
Not specifying the area or location it covers	They work 15 h for 49 pounds

accuracy. It is easy to publish and suddenly reaches millions. This news which is prepared irresponsibly can have serious results both socially and economically when considering how quickly they spread over the internet. For example, a news report, which consisted of the deception in 2008, which claimed that Apple CEO Steve Jobs had a serious heart attack, led to a 10% decline in the company's share price.

In our study, the data was gathered from news websites of media organizations and Twitter to detect this type of news. The data were then tagged as Clickbait or non-Clickbait and compared with TF-IDF in terms of frequency. Afterward, an incoming headline was classified in terms of the reliability index. The accuracy rate of our method was 86.5%.

The rest of the study consists of the following parts. In Sect. 2, related works conducted on the subject are mentioned. In Sect. 3, the data gathered for the Clickbait test and the used filters are described. The adopted methodology and the algorithm are explained in detail. The applications of the test data on the proposed method and the experimental results are analyzed. Finally, the conclusions of the study are presented in Sect. 4.

2 Literature Review

Many studies [2–4] were conducted by observing the behaviors of internet users and making inferences. In their research, Hermann et al. gathered data by using an IP based data gathering method and processed this data by using semi-supervised machine learning method, determining that 87% of the users use the internet with a certain pattern. Müngen and Kaya [5] suggested the activity recommendation system on social networks by using patterns.

SemanticWeb is meaningful data extraction from the web and has been frequently done for the last 15 years [6]. Toma et al. [7] used big data methods to perform SemanticWeb operations. One of the most essential areas of SemanticWeb is natural language processing (NLP) [8–11]. In general, Clickbait news has been detected with NLP methods. One of the first studies on Clickbait was done by Chakraborty et al. [12]. In this study, 6 different news sources were examined. 15,000 news were collected from these news sources. While 7,500 of these were labeled as true news, the remaining 7,500 of news were labeled as Clickbait news.

A learning Clickbait detection method has been created that accepts this news as test data. In this way, it is aimed to detect Clickbait news from the news sites on the internet and to show the news contents without having to click on the news link. With an extension created for a web browser, the news was automatically checked, and news content was shown without clicking.

The study of Potthast et al. [13] is one of the earliest studies trying to identify the Clickbait news on social networks. At the same time, this study is the first Clickbait detection study using machine training method. In the study, the dataset of 2992 Twitter tweets from the top 20 Twitter accounts of 2014 has been prepared. This dataset is also the first Twitter dataset of Clickbait. It includes well-known news publishers such as BBC News, The New York Times, CNN, The Guardian, The Wall Street Journal, and The Washington Post. Mashable is also online. In addition, there are well-known publications such as Business Insider, Huffington Post, and BuzzFeed. 767 of 2992 Tweet has been selected as Clickbait. The majority opinion has been taken into account when deciding whether a tweet is clickbait.

The Clickbait detection model is based on 215 features. Features are divided into three categories: (1) teaser message, (2) linked web page, and (3) meta-data. In the study, it is calculated the correlations according to the features such as retweet, like which are special to twitter and tried to find the post which is Clickbait on Twitter. Using three known learning algorithms, logistic regression (LR), naive Bayes (NB) and random forest (RF), the performance of the study is compared. This study aims to provide a way for consumers and social networks so that their content publishers cannot benefit from Clickbait by filtering related messages.

A competition was organized in 2017 under the leadership of Martin Potthast and his colleague [14], who were working on Twitter. It was asked to present their approach to determining the Clickbait. The dataset to be used by the participants in their studies was also shared on their official sites but it was stated that they can use their own datasets if requested. The criteria of success were also published on the website. At the end of the determined period, winners in a workshop at Bauhaus University in Weimar, Germany were determined and published according to their values from MSE, F1 Score, Precision, Recall, Accuracy, Runtime and many other aspects [15]. Out of the 16 results, 8 were included in the source code and 10 in the format of the article [14]. In August 2018, Potthast, together with his colleagues and a few new people, published another comprehensive study on the mission and the prepared data set [16].

Anand et al. [17] tried to detect Clickbait news using Recurrent Neural Networks (RNN) by taking the dataset prepared by Chakraborty et al. [12]. They rely on existing methods for the detection of Clickbait, based on NLP (Natural Linguistic Programming) toolkits and language-specific words. But, as Potthast uses, they argue that only a method with Twitter metadata can only be applied on Twitter. In the study, it is proposed to use distributed word placements to capture linguistic and semantic features and character placements to capture morphologic features in the RNN model. In the study, an architecture consisting of a Bi-Directional RNN is used, and this RNN output is shown in Fig. 1, which is designed to be of fixed size with the representation of the input. The authors have matched high-

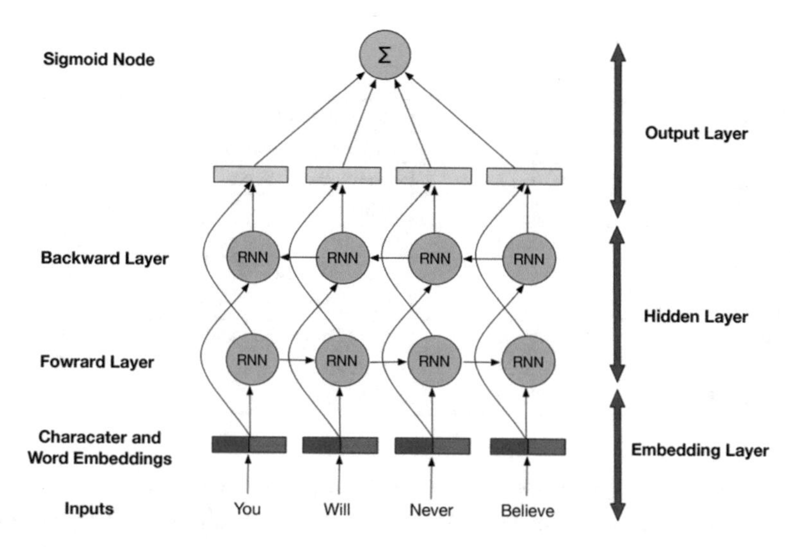

Fig. 1 BiDirectional RNN architecture for detecting Clickbaits [17]

dimensional real-valued vectors with words to capture the hidden semantic and syntactic properties of the words. In the study, 300-dimension word2vec embedding trained using Continuous Bag of Words architecture on approximately 100,000 words was used from Google news data set.

Although Clickbait is not a new problem, it is seen as a big problem with the rapid development of technology and the digital world. Therefore, studies have increased in that subject. Although many studies have been done to prevent this situation, almost all of them have approached as English. This is a shortcoming. Therefore, some researchers have tried a language-independent approach in their studies. As in the study conducted by Hardalov et al. in 2016 [18], Bulgarian news were used in Thai in May 2018 by Wongsap et al. [19], and by Klairith and Tanachutiwat in July 2018 [20], researchers Junfeng Fu, Liang Liang, Xin Zhou, and Jinkun Zheng [21], who wanted to work on a different language from these three languages and determined whether or not the Clickbait would be identified in a much more difficult alphabet, studied Chinese. Hardalov et al. [18] used the features in Table 2. The properties in the table are given in order of importance.

Chesney et al. [22], 2017 EMNLP Workshop published in the study, a study to determine whether the Clickbait, Natural Language Processing methods are not sufficient, argued that the content should be examined semantically.

About this, Blom and Hensen [23] published a study in 2015, Journal of Pragmatics. Blom and Hensen conducted a study on the news headlines collected from 10 news headlines published in Denmark on the 100,000 news headlines. They proved that the headlines of the Clickbait showed decisive features in terms of form, and they referred to it as referrals. In a recent study conducted by Rony et al. [24], 167 million Facebook shares belonging to 153 media institution accounts were analyzed. They trained their systems with this dataset and constructed their

Table 2 Hardalov's property table [18]

Features	Score	Features	Score
doubleQuotes	0.7911	singularPronouns	0.5286
upperCaseCount	0.7748	thirdPersonPronouns	0.5273
lowerUpperCase	0.7717	negativeWordsScore	0.5206
rstUpperCase	0.7708	hashtags	0.4998
pluralPronouns	0.6558	urls	0.4987
rstPersonPronouns	0.6346	positiveWordsScore	0.4910
allUpperCaseCount	0.6282	singleQuotes	0.4884
negativeWords	0.5944	secondPersonPronouns	0.4408
positiveWords	0.5834	questionMarks	0.4407
tokensCount	0.5779	exclMarks	0.3160

methods using distributed sub-word embeddings instead of bag-of-word because of the near-time success of deep learning methods in text classification. They achieved 98.3% accuracy.

The subject of text summary is an implementation popularized by the widespread use of the internet and the increase in the number of readers of comprehensive knowledge and reports. Generally, text summary attempts to find the most attractive areas of a long text and combine these areas in an attempt to conduct the main idea to the reader. Patil et al. [25] conducted summaries by using text feature extraction based on ontology method. Lie et al. [26] used Integer Liner Programming (ILP) in their text summary study and the feature that makes this study stand out from the other is the fact that they included the relation of the images with the text in the method. Ramesh et al. [27] conducted a text summary by using statistical information. Gupta et al. [28] extracted text summaries by using shallow linguistic features. Filho et al. [29] in their study, though they used similar methods to their equivalents, studied scientific texts and published their results comparatively.

3 Clickbait Detection Model

3.1 Data Collection, Storage and Filtering

The first stage of the study has been the detection of the most or least Clickbait news sites and the Twitter accounts of these news sites. At this stage, this is valid for almost all of the contents of the news sites and Twitter accounts which broadcasts in Turkey.. However, we found that some pages contain more or less clickbait news than others. As a result of this stage, organizations such as BBC Turkish and Anadolu Agency are less likely to clickbait, but news sources such as Hürriyet, Vatan and Sabah Newspapers are seen to be quite common in the news published.

Table 3 Count and size of collected data

Type of data	Count	Size (indexed size) (mb)	Clickbait sample	Non-Clickbait sample
Raw data (Twitter)	2670	54	1326	1444
Filtered data (Twitter)	2000	45	1000	1000
Raw data (News sources)	2389	284	1214	1175
Filtered data (News sources)	2000	245	1000	1000

For this reason, these publications and social media accounts of these publications were selected as datasets.

A web spider in Java has been written and news headlines and summaries have been collected at certain intervals to collect news from the newspapers and prepare the news dataset. In terms of the variety and reliability of the news, the spider was run at least 12-hintervals on the same day. The duration of each run was set so that the number of titles collected during that period does not exceed 300. While this number was 2670, the web spider was stopped. Among these titles, 1000 clickbait and 1000 non-clickbait news headers were labeled. The Twitter API was used for data to be collected from Twitter. Again in terms of data reliability and variety, the API was run at regular 12-h intervals and was set to not exceed 300 tweets per cycle. While the number of data collected was 2389, the API was stopped. Twitter posts were labeled as clickbait and non-clickbait tweets. The raw and filtered amounts of the collected data were presented in Table 3.

The texts collected from the websites were removed from their HTML with JSOUP Java library. Additionally, the collected texts were recorded by separating them into headlines, subheadings, and content. The gathered data was first filtered according to categories and only the matters that content politics and the agenda were chosen. Furthermore, rather short tweets and news which were published by shortening due to the media extension were filtered and excluded in the data cluster. The data was stored in ElasticSearch, which is a NoSQL database, due to its convenience for large data processing methods and can reliable examination of the data at big sizes.

3.2 Feature Selection

For the features of news headlines, Hardalov et al. [30] identified the own features at their work. In these studies, the selection of the features was done meticulously, and the determined features were given according to the order of importance, so it was deemed suitable for use in the study. The features that Hardalov et al. identified in their research are given in Table 4, together with their location and significance. These properties represent the content semantically and syntactically. Both the number of upper case letters included are taken into account, and the words used are checked, whether they are semantically positive or negative. Even pronouns are

Table 4 Features ranked by the Lvq importance metric [18]

Features	Rank in property list	Importance
Doublequotes	16	0.7911
Uppercasecount	4	0.7748
Loweruppercase	5	0.7717
Rstuppercase	3	0.7708
Pluralpronouns	6	0.6558
Rstpersonpronouns	8	0.6346
Alluppercasecount	2	0.6282
Negativewords	18	0.5944
Positivewords	17	0.5834
Tokenscount	1	0.5779
Singularpronouns	7	0.5286
Thirdpersonpronouns	10	0.5273
Negativewordsscore	20	0.5206
Hashtags	14	0.4998
Urls	11	0.4987
Positivewordsscore	17	0.4910
Singlequotes	15	0.4884
Secondpersonpronouns	9	0.4408
Questionmarks	13	0.4407
Exclmarks	12	0.3160

considered as singular or plural, first or second person pronouns. Also used in text punctuation marks and hashtag (#) are also important. As we mentioned before, because clickbait titles are more abrasive, punctuation is used more often than non-clickbait headings.

Clickbait title is evaluated with this feature table before it is calculated with TF-IDF. This is a preliminary preparation for frequency measurement with TF-IDF. In this way, errors may occur before another phase is passed. For example, before the height of frequency, the number of upper/lower case letters, URLs, hashtags, positive/negative word numbers, and their scores, number of a question mark and single/double quotes and tokens number are determined.

The features to be extracted from Twitter posts, it has been determined from Potthast et al. studies. Potthast et al. [13] have selected 215 features in three titles: Teaser message, Linked web page and Meta information. The first 203 of these features were selected based on the contents of message, the features between 204 and 211 were used on the linked web page, and the features between 212 and 215 were selected based on the metadata of the message. Whether these features are above 0.5 in according to metric values has been examined. We selected the first 203 properties according to the order of importance of Potthast and selected features with a significance value of over 0.5. All four meta-data properties are also considered.

In our study, we used the first nineteen features related to the Teaser message and the last four properties related to the metadata because we checked the content and

Table 5 Clickbait model values for teaser message

Feature	Precision			Recall			ROC-AUC		
	LR	NB	RF	LR	NB	RF	LR	NB	RF
Meta information	0.62	0.74	0.74	0.55	0.72	0.75	0.54	0.74	0.77
Sender name	0.65	0.65	0.65	0.60	0.60	0.60	0.67	0.67	0.67
Has media attachment	0.55	0.55	0.55	0.53	0.53	0.53	0.47	0.47	0.47
Is reetweet	0.60	0.60	0.60	0.39	0.39	0.39	0.51	0.51	0.51
Part of day as per server time	0.60	0.60	0.60	0.53	0.53	0.53	0.51	0.51	0.51

Table 6 Clickbait model values for meta information

Feature	Precision			Recall			ROC-AUC		
	LR	NB	RF	LR	NB	RF	LR	NB	RF
Teaser message	0.60	0.74	0.71	0.55	0.72	0.73	0.54	0.74	0.73
character 1-g	0.71	0.68	0.71	0.65	0.56	0.71	0.72	0.68	0.71
character 2-g	0.64	0.73	0.71	0.60	0.70	0.72	0.60	0.75	0.74
character 3-g	0.63	0.74	0.74	0.58	0.74	0.75	0.61	0.76	0.77
word 1-g	0.70	0.74	0.72	0.66	0.66	0.71	0.70	0.76	0.77
word 2-g	0.64	0.63	0.61	0.68	0.45	0.46	0.58	0.58	0.55
word 3-g	0.55	0.55	0.55	0.69	0.69	0.69	0.50	0.50	0.50
hashtags	0.64	0.65	0.65	0.32	0.32	0.32	0.50	0.49	0.50
@ mantions	0.71	0.72	0.71	0.37	0.37	0.37	0.53	0.53	0.53
Image tags as per Imagga [13]	0.64	0.64	0.64	0.57	0.57	0.57	0.58	0.58	0.58
Sentiment polarity (Stanford NLP)	0.63	0.63	0.61	0.54	0.55	0.54	0.59	0.59	0.56
Readability (Flesch-Kincaid)	0.67	0.67	0.60	0.59	0.62	0.48	0.65	0.65	0.57
Stop words-to-words ratio	0.09	0.09	0.49	0.30	0.30	0.70	0.50	0.50	0.50
Easy words-to-words ratio	0.57	0.57	0.55	0.48	0.59	0.47	0.50	0.48	0.47
Has abbreviations	0.63	0.64	0.63	0.42	0.37	0.42	0.54	0.54	0.54
Number of dots	0.72	0.72	0.72	0.72	0.72	0.72	0.55	0.55	0.55
Length of longest word	0.62	0.61	0.60	0.49	0.55	0.44	0.57	0.57	0.55
Mean word length	0.58	0.57	0.61	0.51	0.56	0.56	0.50	0.48	0.54
Length of characters	0.67	0.64	0.64	0.59	0.61	0.56	0.62	0.62	0.58

metadata of the message. These properties are given in Tables 5 and 6, according to the values of Precision, Recall, and ROC-AUC determined by three different algorithms.

4 Methodology

In the gathered training data, word count is conducted on the Clickbait ones, calculating the frequency of occurrence by removing the most frequently used words. The same procedure is conducted for the non-Clickbait data, and the most frequently used words and their frequencies are determined. In both lists, because

Table 7 Comparison results

Algorithm	Precision	Recall	Accuracy	F-Measure
Our method	0.899	0.941	0.865	0.920
Recurrent neural network	0.789	0.826	0.738	0.807
Logistic regression (LR)	0.722	0.783	0.664	0.751
Naive Bayes (NB)	0.736	0.771	0.667	0.753
Random forest (RF)	0.713	0.756	0.643	0.734
Feature based	0.698	0.750	0.641	0.723

the common words and their frequencies cannot be used as a distinguishing factor, these types of words are excluded from the lists.

Following the exclusion of the non-distinguishing words, the remaining list and the two lists to be investigated in terms of being a Clickbait are compared to their frequencies by using TF-IDF. Classification is done according to the confidence index resulting from this comparison, and it is decided whether or not the incoming data is Clickbait. If the accuracy index is higher than 0.08 in one side and lower than 0.02 on the other, the Clickbait detection process is completed for that document. In other words, if the news headline's or the subheading's frequency consistency with the Clickbait word list is higher than 0.08 and lower than 0.02 compared to the non-Clickbait word list, this headline is a Clickbait headline. Conversely, if the frequency match is less than 0.02 with the specified Clickbait word list and the frequency match with a non-Clickbait word list is higher than 0.08, the incoming data is not Clickbait. The results of the previous algorithms and the novel method we developed were compared in terms of Precision, Recall, Accuracy, and F-measure metrics. Comparison results were presented in Table 7.

5 Summarization

It has been tried to create a summary by looking at the details of the titles identified as Clickbait in this section. Summarization was done based on the TextRank algorithm, an unsupervised method based on keyphrase proposed by Rada Mihalcea and Paul Tarau [31]. The main reason for using this method is that the news has different content and type, and it cannot be processed by supervised learning. Besides, the detection of the most important sentences in the news is sufficient to express the news. In other words, TextRank is a suitable algorithm for such applications because it calculates and sorts on the text based on the information obtained from the entire text. Summarizing is scoring and ranking all the sentences.

To apply TextRank, a graph associated with the text must first be created. Each sentence in the text is added as a node to the graph. By looking at the similarity between the two sentences, the link between the related nodes is created. For example; a sentence that deals with some concepts in a text refer to other sentences that take the same concepts in the text. A link can be created between these two sentences that share common content. The created graph shows the power of the links between the various sentence pairs in the text. For this reason, the text is converted to a weighted graph. TextRank is very similar to the PageRank algorithm with this feature. PageRank algorithm basically shows the site that has the most links between the sites, and TextRank similarly highlights the sentence that has the most links between sentences.

One of the essential steps of the study is to distinguish between the suffixes of words and their similarities, rather than using a word-based approach. The study is also used in the scoring step for summarizing the similarities of words with the cosine similarity. After all these steps, the sentences are sorted according to their scores. The sentences with the highest scores are selected as the summary.

In our study, several algorithms were tested and used in the summarization section. Finally, it is concluded that the TextRank algorithm is the most appropriate. This algorithm is integrated with the JAVA language for our study. In this selected algorithm, some corrections were made for the Turkish language. Authors and researchers who used the algorithm to decompile the authors' algorithms used English to regex the sentences. So, first, the regex was written in Turkish, so that the mistakes about the separation of words were corrected.

The TextRank algorithm works like processing on a word and tags by type of words to improve success. In other words, this algorithm makes semantic analysis word by word. The process of labeling word types was done in English and Turkish. TextRank algorithm was used for English words in the study. The study was conducted again with JAVA language by using the open source Zemberek-NLP [32] library for the labeling of Turkish words. Although the Turkish and English languages are in the Latin languages, because the classification style for words is different, the labels of the Turkish and English words are matched by similarity.

Another factor that increases the success of the study is that the words which are not necessary for the sentence such as preposition and conjuncture which do not change the meaning are accepted as stop words. They are removed from the system, and the algorithms are applied according to this situation. In this way, the success rate of the summarization function is increased.

6 Conclusion

Whether the news is Clickbait or not is found by comparing the frequency of the headline with the frequency of the words in the test data. On the other hand, adding news that is detected as Clickbait or not Clickbait for training data can increase the success rate and the discovered Clickbait words can be recognized automatically. Training data can be developed with machine learning methods. Since the dataset in this study is Turkish in general, stop words, and Clickbait words are Turkish too. On the other hand, depending on the language the user speaks, the region or even the user audience, adding or updating the words for different languages can increase the success rate. In social networks that have limited character like Twitter, it is difficult to detect Clickbait, but the success rate can be increased by adding the features such as retweet number, like the number in the social network as parameters to the algorithm.

In the future, increasing the diversity of the dataset is one of the first factors that will extend the study and improve the accuracy rate. Not only the social networks and websites but also the other news content like mobile phone notifications or words on television subtitles can be added to the dataset. Moreover, increasing the number of languages in the dataset can be another study related to the dataset.

The inclusion of other machine learning methods and deep learning methods in this study will be an essential step to increase the success rate and make a more reliable system. The results of the survey will be evaluated comparatively with both our method and previously tested methods to find the ideal solution for Clickbait detection.

Acknowledgement This study was supported Firat University, Scientific Research Projects Office under Grant No MF.18.07.

References

1. Digital in 2017: global overview - we are social (2017), https://wearesocial.com/special-reports/digital-in-2017-global-overview. Accessed 20 Dec 2017
2. T. Leppänen, I.S.J.Y. Milara, J. Kataja, J. Riekki, Enabling user-centered interactions in the Internet of Things, in *2016 IEEE International Conference on Systems, Man, and Cybernetics (SMC)* (IEEE, 2016)
3. H. Li, G. Ye, X. Liu, F. Zhao, D. Wu, X. Lin, URLSight: profiling mobile users via large-scale internet metadata analytics, in *Trustcom/BigDataSE/ISPA, 2016* (IEEE, 2016)
4. A. Teichman, S. Thrun, Tracking-based semi-supervised learning. Int. J. Rob. Res. **31**(7), 804–818 (2012)
5. A.A. Müngen, M. Kaya, A novel method for event recommendation in meetup, in *2017 IEEE/ACM International Conference on Advances in Social Networks Analysis and Mining (ASONAM)* (IEEE, 2017)
6. T. Berners-Lee, J. Hendler, O. Lassila, The semantic web. Sci. Am. **284**(5), 34–43 (2001)
7. I. Toma, D. Roman, K. Iqbal, D. Fensel, S. Decker, J. Hofer, Towards semanticweb services in grid environments, in *First International Conference on Semantics, Knowledge and Grid, 2005. SKG '05* (IEEE, 2005)
8. E.D. Liddy, E. Hovy, J. Lin, J. Prager, D. Radev, L. Vanderwende, R. Weischedel, Natural language processing, in *b*, 2nd edn., (Marcel Decker, Inc., New York, 2003), pp. 2126–2136
9. L. Greco, P. Ritrovato, A. Saggese, M. Vento, Abnormal event recognition: a hybrid approach using semanticweb technologies, in *2016 IEEE Conference on Computer Vision and Pattern Recognition Workshops (CVPRW)* (IEEE, 2016)
10. D. Skoutas, A. Simitsis, T. Sellis, A ranking mechanism for semanticweb service discovery, in *2007 IEEE Congress on Services* (IEEE, 2017)
11. A.A. Müngen, M. Kaya, Mining quad closure patterns in Instagram, in *2016 IEEE/ACM International Conference on Advances in Social Networks Analysis and Mining (ASONAM)* (IEEE, 2016)
12. A. Chakraborty, B. Paranjape, S. Kakarla, N. Ganguly, Stop clickbait: detecting and preventing clickbaits in online news media, in *IEEE/ACM International Conference on Advances in Social Networks Analysis and Mining* (IEEE, 2016)

13. M. Potthast, S. Köpsel, S. Benno, M. Hagen, Clickbait detection, in *38th European Conference on IR Research* (ECIR, 2016), pp. 810–817
14. Clickbait Challenge 2017 (2017), https://www.clickbait-challenge.org
15. Clickbait Challenge 2017 Dataset (2017), https://www.tira.io/task/clickbait-detection/dataset/clickbait17-test-170720/
16. M. Potthast et al., Crowdsourcing a large corpus of clickbait on Twitter, in *Proceedings of 27th International Conference on Computational Linguistics, Santa Fe, New Mexico, USA* (Association for Computational Linguistics, 2018), pp. 1498–1507
17. A. Anand, T. Chakraborty, N. Park, We used neural networks to detect clickbaits: you won't believe what happened next, in *39th European Conference on IR Research* (ECIR, 2017), pp. 541–547
18. M. Hardalov, I. Koychev, P. Nakov, In search of credible news, in *International Conference on Artificial Intelligence: Methodology, Systems, and Applications*, Lecture Notes in Computer Science (including subseries Lecture Notes in Artificial Intelligence and Lecture Notes in Bioinformatics) (Springer, Cham, 2016)
19. N. Wongsap, L. Lou, S. Jumun, T. Prapphan, S. Kongyoung, N. Kaothanthong, Thai clickbait headline news classification and its characteristic, in *2018 International Conference on Embedded Systems and Intelligent Technology and International Conference on Information and Communication Technology for Embedded Systems, ICESIT-ICICTES 2018* (2018)
20. P. Klairith, S. Tanachutiwat, Thai clickbait detection algorithms using natural language processing with machine learning techniques, in *International Conference on Engineering, Applied Sciences, and Technology (ICEAST)* (IEEE, 2018)
21. J. Fu, L. Liang, X. Zhou, J. Zheng, A convolutional neural network for clickbait detection, in *Proceedings - 2017 4th International Conference on Information Science and Control Engineering, ICISCE 2017* (IEEE, 2017)
22. S. Chesney, M. Liakata, M. Poesio, M. Purver, Incongruent headlines: yet another way to mislead your readers, in *Proceedings of the 2017 EMNLP Workshop: Natural Language Processing Meets Journalism*. Copenhagen, Denmark (2017), pp. 56–61
23. J.N. Blom, K.R. Hansen, Click bait: forward-reference as lure in online news headlines. J. Pragmat. **76**, 87 (2015)
24. M.U. Rony, N. Hassan, M. Yousuf, Diving deep into clickbaits: who use them to what extents in which topics with what effects? in *2017 IEEE/ACM International Conference on Advances in Social Networks Analysis and Mining (ASONAM)* (ACM, 2017)
25. A.R. Patil, A.A. Manjrekar, A novel method to summarize and retrieve text documents using text feature extraction based on ontology, in *IEEE International Conference on Recent Trends in Electronics, Information & Communication Technology (RTEICT)* (IEEE, 2016)
26. W. Li, H. Zhuge, Summarising news with texts and pictures, in *2014 10th International Conference on Semantics, Knowledge and Grids (SKG)* (IEEE, 2014)
27. A. Ramesh, K.G. Srinivasa, N. Pramod, SentenceRank—a graph based approach to summarize text, in *2014 Fifth International Conference on the Applications of Digital Information and Web Technologies (ICADIWT)* (IEEE, 2014)
28. P. Gupta, V.S. Pendluri, I. Vats, Summarizing text by ranking text units according to shallow linguistic features, in *13th International Conference on Advanced Communication Technology* (IEEE, 2011), pp. 1620–1625.
29. P.P.B. Filho, T.A.S. Pardo, M.D.G.V. Nunes, Summarizing scientific texts: experiments with extractive summarizers, in *Proceedings of the 7th International Conference on Intelligent Systems Design and Applications, ISDA 2007* (IEEE, 2007), pp. 520–524
30. M. Hardalov, I. Koychev, S. Kliment, In search of credible news, in *International Conference on Artificial Intelligence: Methodology, Systems, and Applications* (Springer, Cham, 2016), pp. 172–180
31. R. Mihalcea, P. Tarau, TextRank: bringing order into texts, in *Proceedings of the 2004 Conference on Empirical Methods in Natural Language Processing*, vol 85 (2004), pp. 404–411
32. Zemberek-NLP, https://github.com/ahmetaa/zemberek-nlp. Accessed 11 Dec 2017

A Model-Based Approach for Mining Anomalous Nodes in Networks

Mohamed Bouguessa

1 Introduction

1.1 Context

Information networks have become an emerging model for representing various interaction scenarios among entities of real world systems. With nodes representing entities and edges reflecting their interactions, many different social, biological and technological systems can be described by means of a network. Drawing on analytic tools from graph theory and data mining, the analysis of information networks is the subject of an increasing attention. Networks analysis include community detection [1], link prediction [2], anomalous nodes identification [3], tracking evolving community structures [4], etc. In this paper, we focus on the problem of anomalous nodes identification.

Mining anomalous nodes is an important network analysis task which finds applications, for example, in the field of fraud detection, network robustness analysis and intrusion detection. Anomaly detection (and removal) can also be implemented as a pre-processing step prior to the application of an advanced network analysis approach. An anomaly could be broadly defined as "an observation which appears to be inconsistent with the reminder of the data set" [3, 5]. The notion of inconsistency here remain general as its differ from one application to another. This makes the problem of anomaly detection an open-ended one [3] with many definitions of an anomaly depending on the application domain or the problem under investigation. In this study, we are interested by handling anomalous nodes in static

M. Bouguessa (✉)
Department of Computer Science, University of Quebec at Montreal, Montreal, QC, Canada
e-mail: bouguessa.mohamed@uqam.ca

M. Kaya et al. (eds.), *Putting Social Media and Networking Data in Practice for Education, Planning, Prediction and Recommendation*, Lecture Notes in Social Networks, https://doi.org/10.1007/978-3-030-33698-1_12

networks. That is, networks that are represented as plain graphs in which two nodes are related by one unlabeled links. The problem of mining anomalies in dynamic or attributed networks is far beyond the scope of this study. Hence, in the reminder of this paper, unless otherwise specified, the term network refers to a static graph in which two nodes are related by unweighted and undirected link.

Akoglu et al. [3], in their through survey paper, suggested a general definition of the network anomaly detection problem as the task of identifying nodes that are different or deviate significantly from the patterns observed in the graph. In simple terms, anomalies (also referred to as outliers, exceptions, abnormal, etc.) correspond to isolated/atypical nodes that do not belong to any identifiable dense region/community structure in the network. As suggested in [3], this may include bridge nodes that do not directly belong to one particular community or nodes that have many cross connections to multiple different communities. To illustrate, as can be seen from Fig. 1, we generated a network of 135 nodes. 92 nodes form 11 communities, while the remaining 43 nodes are anomalies. Normal nodes are those nodes that are closely connected to each other and thus form community structures which are delimited by circles as illustrated by Fig. 1. In this figure, nodes outside these circles correspond to anomalies. As we can see, we have different topological configuration of anomalies. For example: the node having many connections across communities $C1$, $C3$, $C4$ and $C5$; the bridging nodes between communities $C2$ and $C3$; nodes forming isolated chain of two to four connected nodes; nodes forming circular chain of more than four connected nodes; and peripheral nodes that are weakly connected to members of a community. Our goal is to devise a principled

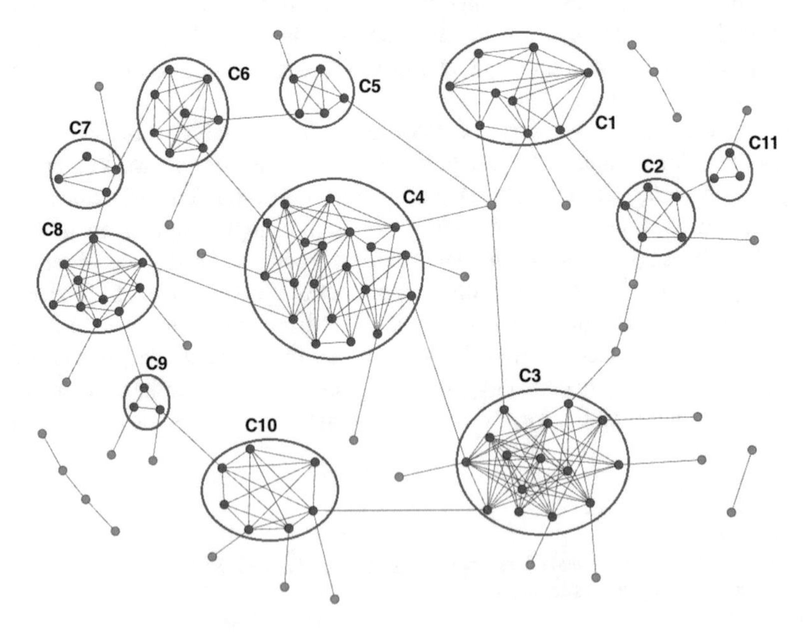

Fig. 1 A network with communities and anomalies

approach to automatically identify such kind of anomalous nodes without asking the user to set any detection threshold or specifying the target number of anomalous nodes to identify from a ranked list.

1.2 Background Information

As just discussed, in this study we focus on the problem of detecting anomalous nodes in static graphs. More details and surveys on noisy edges detection and anomalous nodes identification in dynamic and attributed networks can found, for example, in [3, 5–7]. In the follows, we provide a high-level description of some relevant approaches related to this work. In the current literature, a number of approaches for identifying anomalous nodes in static graph data have been proposed. In this context, as suggested in [3], anomaly detection approaches exploit the topological structure of the network to spot anomalies that deviate significantly from structural patterns (community structures for example) observed in the graph. Such approaches can be broadly classified as feature-based, community-based, and scoring-based [3]. Feature-based approaches associate to each node a feature vector which encompass several metrics such as betweenness, closeness, local clustering coefficient, node degree, to name but a few. Depending on the application domain, other features extracted from additional information sources can be used together with the topological features calculated from the graph. These features are then used as input for a classical outlier detection algorithm so that nodes associated with feature vectors identified as outliers (in the feature space) will be declared as anomalies. OddBall [8] is an example of a feature-based approach which is based on the assumption that a node is anomalous if its neighborhood differs significantly from those of others. The notions of ego (a unique node) and egonet (the subgraph of the neighboring nodes of the ego) have been introduced based on which four egonet-based features have been extracted. The LOF algorithm [9] is then used to obtain an outlierness score of each feature vector. Next, in order to identify various kind of outliers, this LOF score is summed up to another distance-to-fitting-line-based anomaly sore. The final outlier scores are used to rank nodes expecting anomalous nodes to come first.

Community-based approaches assume that anomalies are nodes that do not cluster well. As suggested in [3], these approaches attempt to identify dense regions in the graph (clusters or communities) and spot nodes that are weakly connected to communities (this include nodes having many connections across several communities and bridging nodes). For example, SCAN [10] exploit the neighborhood of nodes in such a way that nodes sharing may neighbors are grouped together. Any nodes that do not belong to the identified clusters are flagged as outliers. SCAN requires a user similarity threshold to be specified in order to perform outliers' detection. gSkeletonClu [11] is a density-based algorithm which is inspired from the DBSCAN algorithm to detect dense regions in the graph and outliers. The main idea is, given a minimum similarity ϵ and a minimum cluster size

μ, each node in a cluster must have at least μ neighbors whose structural similarities are at least ϵ. In contrast to SCAN, gSkeletonClu is able to automatically identify the minimum similarity threshold but requires the minimum cluster size as input parameter. NNrMF [12] another community-based algorithm which relies on matrix factorization to detect community structures. The top k edges of the residual matrix are considered to spot anomalies. NNrMF requires the rank size r to be specified to estimate the residual matrix. Finally, scoring-based approaches calculate some kind of score per node which serves as a measure of the degree of anomalousness in such way that central and informative nodes receive high score values compared to marginal or atypical nodes. PageRank-like measures [13], for example, can be used to estimate the anomaly scores which can be used in ranking nodes such that top k nodes are considered as outliers.

1.3 Motivations

Overall, the problem of anomaly detection in static networks (plain graphs) has been well studied for which a number of approaches have been proposed in the literature. In spite of these existing techniques, the problem of the automatic detection of anomalous nodes continues to pose a challenge to existing algorithms in various ways. For instance, the output of a number existing approaches [8, 13] is a ranked list of nodes that represent the degree of anomalousness of each node. The top k nodes with the highest score values are considered as anomalies. A major problem for such an approach is determining how many data points should be chosen as anomalies from a ranked list. In general, the number of outliers to be selected from a ranked list is specified in and ad hoc manner. Here, it is clear that with such an unprincipled approach, it is impossible to be objective or consistent. Other existing approaches [10] suffer from their dependency to some critical user parameters, such as a similarity threshold, which may be difficult to tune. Here, it is clear that incorrect settings may cause an algorithm to fail in finding the "true" anomalies. The analyst needs thus to run the anomaly detection procedure many times with different input parameters to get a feel for which result might be more reasonable. This in turn makes the uses of parameter-laden anomaly detection methods tricky in practical applications.

Another limitation of existing models is who to combine various anomaly scoring metrics to automatically detect anomalous nodes. As mentioned in the previous subsection, OddBall [8] combine LOF outliers scores with distance-to-fitting-line-based anomaly sore by a simple addition to identify various kind of outliers. Such naive approach, that is, the summation of different scores, appears to be less appropriate especially in some real application when anomaly scores differ in their range and their meaning. Here, we believe that a notable advance than "yet another anomaly detection algorithm" is a principled way to combine between scores that help to distinguish normal form atypical nodes. This study proposes an approach towards this goal.

1.4 Contributions

Starting from the assumption that normal nodes tend to be densely connected compared to anomalous nodes that are sparsely distributed across the network; evaluating the connectivity patterns of nodes is a plausible approach to distinguish between normal and anomalous nodes. In this setting, node similarity measures can be used. The main question here, what is the appropriate measure that we should consider to effectively quantifies nodes connectivity under various network characteristics, so that we can efficiently spot anomalous nodes. This remains an open question as, depending on the application at hand, nodes connectivity may significantly differ from one region of the network to another and, in this setting, it is hard to determine what is the appropriate metric. In fact, node similarity metrics are primarily developed for a specific task and their effectiveness vary among networks. In many cases, a metric is effective on handling anomalies in networks with certain properties while delivering poor performances on even small changes in the underlying network characteristics.

In order to yield a high anomaly detection accuracy, an efficient approach is to combine the results of multiple node similarity metrics. To this end, we first associate to each node a neighborhood cohesiveness feature vector, such that each element of this vector corresponds to a score quantifying node's neighborhood connectivity, as estimated based on a specific similarity measure. Next, we propose a statistical framework, based on the Dirichlet mixture, in order to model the estimated neighborhood feature vectors. The probability density function is therefore estimated and the Dirichlet component that corresponds to anomalous nodes is identified.

We propose the use of the Dirichlet distribution because, in contrast to several distributions, it is more flexible and powerful since it permits multiple symmetric and asymmetric modes, it may be skewed to the right, skewed to left or symmetric [14]. This great shape flexibility of the Dirichlet distribution provides a better fitting of the node's feature vectors, which leads, in turn, to a substantially improved modeling accuracy. Our experimental results corroborate our claim. We summarize the significance of this work as follows.

- We propose a principled approach, based on the Dirichlet mixture, to combine the results of several similarity metrics that reflect the neighboring connectivity of a node. By doing so, we aim to take advantage of the collective strength of the metrics and avoid their individual pitfalls.
- The approach that we propose is able to handle various topological structures of anomalies: nodes connected across communities, bridging nodes, extreme peripheral nodes and isolated chain nodes.
- The proposed approach identifies anomalous node in an automatic manner rather than providing a ranked list of nodes and asking the user to specify the number of anomalies to be identified or requiring a detection threshold to detect anomalies.

2 Proposed Approach

In this section, we elaborate an approach that combines various scores reflecting the neighboring connectivity of a node in order to gain an improved anomaly detection effectiveness. To this end, we propose to associate to each node u_i ($i = 1, \ldots, N$), where N denote the number of nodes in the network under investigation, a neighborhood cohesiveness vector $\vec{X}_i = (x_{i1}, \ldots, x_{id})^T$. Each element of the vector quantifies the closeness of a node with respect to its neighbors based on a specific similarity measure S_j, $j = 1, \ldots, d$. The dimensionality d of $\vec{X}_i$ is equal to the number of similarity measures used. Several exiting metrics can be used for this task. Here, we draw the attention of the reader that in this paper, we do not aim at defining new proximity or closeness metrics. This would be far beyond the scope of this study. In this work, we utilize existing measures and we focus on illustrating the suitability of our Dirichlet mixture-based approach in effectively combining the results of various metrics to automatically identify anomalous nodes.

2.1 Neighborhood Cohesiveness Feature Vectors Estimation

Starting from the assumption that normal nodes tend to be closely connected and thus share many neighbors compared to anomalies, we investigate the vicinity of a node to distinguish between "close-by" nodes (normal nodes) from sparse and atypically connected nodes (anomalies). Node similarity metrics can be used for this task. Other topological-based metrics can also be used. However, as observed in our experiments, this will unnecessarily inflate the running time of our approach without gaining any notable improved detection accuracy. For the sake of simplicity, we mainly focus on similarity metrics to evaluate nodes' connectivity by calculating the similarity between each node and its immediate neighboring nodes. This step aims to reveals the connection's strength between pairs of nodes as similarity measures describe how they are tightly connected and how much similar are their neighborhoods.

Let $N(u)$ and $N(v)$ denote, respectively, the set of neighboring nodes of u and v. Among exiting metrics, we elected the use of four alternative similarity measures to evaluate the similarity between u and v.

1. **The number of shared neighbors:** $S_1(u, v) = |N(u) \cap N(v)|$;
2. **The Jaccard Index:** $S_2(u, v) = \frac{|N(u) \cap N(v)|}{|N(u) \cup N(v)|}$;
3. **Meet/Min:** $S_3(u, v) = \frac{|N(u) \cap N(v)|}{min(|N(u)|, |N(v)|)}$;
4. **Geometric:** $S_4(u, v) = \frac{|N(u) \cap N(v)|^2}{(|N(u)||N(v)|)}$.

$S_1(u, v)$ counts the number of shared neighbors between u and v, but ignore the ignore the size of the entire neighborhoods of u and v combined. The Jaccard index, that is, $S_2(u, v)$, can be used to alleviate this drawback since it favors node pairs for which a high percentage of all neighbors are shared. However, in some situations, this measure may not to reflect well the neighbors' connectivity if one of the two connected nodes has a large neighborhood [15]. The Meet/Min coefficient, that is, $S_3(u, v)$, removes this bias at the expense of discarding information about the larger neighborhood size [15]. Finally, the geometric coefficient, that is, $S_4(u, v)$, is a compromise between the Jaccard and Meet/Min. Here, we can claim that the selected four metrics investigate nodes vicinity from different perspective. In order to take advantage of their collective strength and avoid their individual drawbacks, we believe that a thorough combination of the results of these metrics can yields an improved detection accuracy. We show in the follows how to effectively combine the results obtained by these metrics in a probabilistic framework, based on which we can automatically identify anomalous nodes.

In general, the aforementioned similarity measures return high values for strongly connected nodes, that is, nodes sharing a relatively large number of common neighbors. On the other hand, a small value of these metrics indicates that u and v are sparsely connected in the sense that they share very few number of common neighbors. Since anomalous nodes tend to be weakly connected to the remaining nodes of the network, it is clear here that the values returned by similarity measures tend to be very small. So, each S_j provides a relative measure to reflect the connection's strength between nodes. Our objective now, is to exploit these scores to estimate the vector $\vec{X}_i$ for each node u_i so that each element x_{ij} of this vector reflects the strength of connectivity of u_i with the respect to its neighbors $N(u_i)$.

Since anomalous nodes ten to be weakly connected to the remaining nodes of the networks, the values of S_j associated to anomalies tend to be relatively small compared to those of strongly connected nodes (normal nodes). In this setting, calculating the maximum value of each S_j associated to a node u_i with respect to its immediate neighbors $v \in N(u_i)$ is a fair indication of the neighborhood cohesiveness around each node in the network. Specifically, for each node u_i, each element x_{ij} of $\vec{X}_i$ can be estimated as

$$x_{ij} = \max\Big(S_j(u_i, v)\Big), \quad v \in N(u_i) \tag{1}$$

According to (1), nodes that share a large number of common neighbors will get a large value of x_{ij} in comparison to sparsely connected nodes that do not have any (or a very low number of) common neighbors with respect to the rest of the network nodes. In other words, smallest feature values $\{x_{ij}\}$ may potentially correspond to anomalies, while the largest ones represent normal nodes. Let us focus now in exploiting these scores to automatically discriminate anomalous from normal nodes without using any ranking-based methodology nor asking the user to set a detection threshold. To this end, we describe in the follows a probabilistic model which combines the estimated neighborhood cohesiveness feature values $\{x_{ij}\}$ of each vector $\vec{X}_i$ for an improved anomaly detection.

2.2 The Probabilistic Model Framework

The main goal of statistical modeling is to establish a probabilistic model which can characterize the patterns of the observations, capture their underlying distributions, and describe the statistical properties of the source [16]. Mixture models are flexible and powerful probabilistic tools for analyzing data. The approach of mixture model assumes that the observed data is drawn from a mixture of parametric distributions. Several parametric statistical models could be used to describe the statistical properties of $\vec{X}_i$. In this paper we propose to use the Dirichlet mixture model.

We have used the Dirichlet distribution mainly because it permits multiple modes and asymmetry and can thus approximate a wide variety of shapes [14], while several other distributions are not able to do so. For example, the use of the Gaussian distribution appears to be restrictive since this distribution permits symmetric "bell" shape only. However, in many real life applications, the data under investigation are non-Gaussian, that is, skewed data with non-symmetric shapes. In this setting, the use of the popular Gaussian distribution, for example, may lead to inaccurate modeling (e.g. over estimation of the number of components in the mixture, increase of misclassification errors, etc.) because of its symmetric shape restriction [17]. Due to the limitations of the Gaussian distribution, we believe that this distribution could not be used to cluster the nodes' feature vectors into several components. The number of components in the mixture will be over-estimated and the identification of anomalous nodes will be, in turn, not obvious. In this paper we use more flexible and powerful mixture model which can thus approximate a wide variety of shapes. In fact, in contrast to several distributions such as the Gaussian or the Gamma, the Dirichlet distribution is very versatile and it is capable of modelling a variety of uncertainties. The Dirichlet distribution may be L-shaped, U-shaped, J-shaped, skewed to the left, skewed to the right, or symmetric. The shapes of Gaussian, gamma and uniform distributions are special cases of the Dirichlet distribution. Such great shape flexibility enables the Dirichlet distribution to provides an accurate fit of the estimated neighborhood cohesiveness feature vectors.

Specifically, in this work, we assume that $\vec{X}_i$ can be considered as coming from several underlying probability distributions. Each distribution is a component of the Dirichlet mixture model that represents a set of nodes' neighborhood cohesiveness feature vectors which are close one to another, and all the components are combined by a mixture form. The component which contains vectors with the lowest values corresponds to anomalous nodes. Note that, to fit the Dirichlet distribution, we normalize the scores of each node's feature vector in such way that the summation of all the d element of the vector $\vec{X}_i = (x_{i1}, \ldots, x_{id})^T$ is smaller than one. All along of this paper, we only use the normalized values of the nodes' feature vectors $\vec{X}_i$.

Formally, we expect that $\{\vec{X}_i\}$ follow a mixture density of the form

$$F(\vec{X}_i|\alpha, \vec{a}) = \sum_{c=1}^{C} \alpha_c F_c(\vec{X}_i|\vec{a_c}) \tag{2}$$

where F_c is the cth Dirichlet distribution, C denotes the number of components in the mixture, $\vec{a_c} = (a_{c1}, \ldots, a_{c(d+1)})^T$ is the parameter vector of the cth component, and $\alpha = \{\alpha_1, \ldots, \alpha_C\}$ represents the mixing coefficients which are positive and sum to one. The density function of the cth component is given by

$$F_c(\vec{X}_i|\vec{a_c}) = \frac{\Gamma\left(|\vec{a_c}|\right)}{\prod_{j=1}^{d+1} \Gamma(a_{cj})} \prod_{j=1}^{d+1} x_{ij}^{a_{cj}-1} \tag{3}$$

where $|\vec{a_c}| = \sum_{j=1}^{d+1} a_{cj}$ $(a_{cj} > 0)$, $\sum_{j=1}^{d} x_{ij} < 1$ $(0 < x_{ij} < 1)$, $x_{i(d+1)} = 1 - \sum_{j=1}^{d} x_{ij}$, and $\Gamma(.)$ is the gamma function given by $\Gamma(\lambda) = \int_0^\infty \gamma^{\lambda-1} \exp(-\gamma) d\gamma;\ \gamma > 0$.

2.3 Parameters Estimation

The most central task in modeling node's feature vectors with the Dirichlet mixture is parameters estimation. To this end, the maximum likelihood estimation approach can be used to find the parameters of the mixture model. Let $\mathbf{P} = \{\alpha_1, \ldots, \alpha_C, \vec{a_1}, \ldots, \vec{a_C}\}$ denotes the set of unknown parameters of the mixture and $\mathbf{X} = \{\vec{X}_1, \ldots, \vec{X}_N\}$ the set of the normalized nodes' feature vectors. The likelihood function corresponding to C components is defined as

$$Lik(\mathbf{X}|\mathbf{P}) = \prod_{i=1}^{N} \sum_{c=1}^{C} \alpha_c F_c(\vec{X}_i|\vec{a_c}) \tag{4}$$

The maximum likelihood of the mixture parameters can be estimated using the Expectation Maximization (EM) algorithm. Accordingly, we augment the data by introducing a C-dimensional indication vector $\vec{Y}_i = (y_{i1}, \ldots, y_{iC})^T$ for each $\vec{X}_i$. The indication vector $\vec{Y}_i$ has only one element equals 1 and the remaining elements equal 0. If the cth element of $\vec{Y}_i$ equals 1, that is, $y_{ic} = 1$, we assume that $\vec{X}_i$ was generated from the cth component of the mixture. Let $\mathbf{Y} = \{\vec{Y}_1, \ldots, \vec{Y}_N\}$ denotes the set of indication vectors. The likelihood function of the complete data is given by

$$Lik(\mathbf{X}, \mathbf{Y}|\mathbf{P}) = \prod_{i=1}^{N} \prod_{c=1}^{C} \left[\alpha_c F_c(\vec{X}_i|\vec{a_c})\right]^{y_{ic}} \tag{5}$$

Usually, it is more convenient to work with the logarithm of the likelihood function which is equivalent to maximizing the original likelihood function. Consequently, the log-likelihood function is given by

$$\begin{aligned} LogLik(\mathbf{X}, \mathbf{Y}|\mathbf{P}) &= \log(Lik(\mathbf{X}, \mathbf{Y}|\mathbf{P})) \\ &= \sum_{i=1}^{N}\sum_{c=1}^{C} y_{ic}\big[\log(\alpha_c) + \log(F_c(\vec{X}_i|\vec{a}_c))\big] \end{aligned} \tag{6}$$

From this perspective, the EM algorithm can be used to estimate **P**. Specifically, the algorithm iterates between an Expectation step and a Maximization step in order to produce a sequence estimate $\{\hat{\mathbf{P}}\}^{(r)}$, $(r = 0, 1, 2, \ldots)$, where r denotes the current iteration step, until the change in the value of the log-likelihood in (6) is negligible. Details of each step are given in the follows.

In the Expectation step, each latent variable y_{ic} is replaced by its expectation:

$$\hat{y}_{ic}^{(r)} = E[y_{ic}|\vec{X}_i, \mathbf{P}] = \frac{\hat{\alpha}_c^{(r)} F_c(\vec{X}_i|\vec{\hat{a}}_c)}{\sum_{k=1}^{C} \hat{\alpha}_k^{(r)} F_k(\vec{X}_i|\vec{\hat{a}}_k)} \tag{7}$$

In the Maximization step, the set of parameters $\mathbf{P} = \{\alpha_1, \ldots, \alpha_C, \vec{a}_1, \ldots, \vec{a}_C\}$ that maximize the log-likelihood are calculated given the values of $\widehat{y}_{ic}$ estimated in the Expectation step. Specifically, the mixing coefficients are calculated as

$$\hat{\alpha}_c^{(r+1)} = \frac{\sum_{i=1}^{N} \hat{y}_{ic}^{(r)}}{N}, \; c = 1, \ldots, C \tag{8}$$

Let us now focus on estimating the parameters $\{\vec{a}_c = (a_{c1}, \ldots, a_{c(d+1)})^T\}_{(c=1,\ldots,C)}$. The values of $\{\vec{a}_c\}$ that maximize the likelihood can be obtained by taking the derivative of the log-likelihood of the complete data with respect to a_{cj} and setting the gradient equal to zero. We obtain

$$\frac{\partial LogLik(\mathbf{X}, \mathbf{Y}|\mathbf{P})}{\partial a_{cj}} = \sum_{i=1}^{N} \widehat{y}_{ic} \frac{\partial}{\partial a_{cj}} \log(F_c(\vec{X}_i|\vec{a}_c)) = 0 \tag{9}$$

By replacing $F_c(\vec{X}_i|\vec{a}_c)$ by its expression given by (3) in (9) and then computing the derivative with respect to a_{cj}, we obtain

$$\sum_{i=1}^{N} \widehat{y}_{ic}\big[\psi(|\vec{a}_c|) - \psi(a_{cj}) + \log(x_{ij})\big] = 0 \tag{10}$$

where $\psi(.)$ is the digamma function given by $\psi(\lambda) = \frac{\partial log\Gamma(\lambda)}{\partial\lambda}$.

Since the gamma function is defined though an iteration, a closed-form solution to (10) does not exists. Therefore, the values of the parameter vectors $\{\vec{a_c}\}$ can be estimated using the Newton-Raphson method. Specifically, we estimate the $\vec{a_c}$ iteratively:

$$\left[\vec{\hat{a}_c}\right]^{(r+1)} = \left[\vec{\hat{a}_c}\right]^{(r)} - M_c^{-1} \times D_c,\ c = 1, \ldots, C \tag{11}$$

where D_c is the first derivative vector of the complete log-likelihood:

$$D_c = \left(\frac{\partial LogLlik(X, Y|P)}{\partial a_{c1}}, \ldots, \frac{\partial LogLik(X, Y|P)}{\partial a_{c(d+1)}}\right)^T \tag{12}$$

and M is the Hessian matrix. The diagonal elements of M correspond to the second derivative of the complete log-likelihood function while the non-diagonal elements correspond to the mixing derivatives [14]. The second derivative is given by:

$$\frac{\partial^2 LogLik(X, Y|P)}{\partial^2 a_{cj}} = \left(\acute{\psi}\left(|\vec{a}_c|\right) - \acute{\psi}\left(a_{cj}\right)\right) \times \sum_{i=1}^{N} y_{ic},\ j = 1, \ldots, d+1 \tag{13}$$

and the mixed derivative is given by:

$$\frac{\partial^2 LogLik(X, Y|P)}{\partial a_{cj_1} a_{cj_2}} = \left(\acute{\psi}\left(|\vec{a}_c|\right)\right) \times \sum_{i=1}^{N} y_{ic},\ \ j_1 \neq j_2,\ j = 1, \ldots, d+1 \tag{14}$$

where $\acute{\psi}$ (.) is the trigamma function given by $\acute{\psi}(\lambda) = \frac{\acute{\acute{\Gamma}}(\lambda)}{\Gamma(\lambda)} - \left[\frac{\acute{\Gamma}(\lambda)}{\Gamma(\lambda)}\right]^2$. Thus, the Hessian matrix can be defined as follows:

$$M_c = \sum_{i=1}^{N} y_{ic} \times \begin{pmatrix} \acute{\psi}\left(|\vec{a}_c|\right) - \acute{\psi}\left(a_{c1}\right) & \ldots & \acute{\psi}\left(|\vec{a}_c|\right) \\ \vdots & \ddots & \vdots \\ \acute{\psi}\left(|\vec{a}_c|\right) & \ldots & \acute{\psi}\left(|\vec{a}_c|\right) - \acute{\psi}\left(a_{cd+1}\right) \end{pmatrix} \tag{15}$$

The inverse of the Hessian matrix is calculated according to [18] by employing the replace form of the Hessian matrix as follows:

$$M_c = H_c + \delta_c R_c R_c^T \tag{16}$$

where H_c is a diagonal matrix given by:

$$H_c = diag\left[-\sum_{i=1}^{N} y_{ic}\acute{\psi}\left(c_{c1}\right), \ldots, -\sum_{i=1}^{N} y_{ic}\acute{\psi}\left(\alpha_{cd+1}\right)\right] \tag{17}$$

δ_c and R_c^T are given as follows:

$$\delta_c = \sum_{i=1}^{N} y_{ic} \grave{\psi} \left(|\vec{a}_c| \right) \tag{18}$$

$$R_c^T = (r_1, \ldots, r_{d+1}); \;\; r_j = 1, \;\; j = 1, \ldots, d+1 \tag{19}$$

The inverse of the Hessian matrix can now be calculated according to the theorem of matrix inverse [19] as follows:

$$M_c^{-1} = H_c^{-1} + \delta_c^* R_c^{*T} R_c^* \tag{20}$$

where H_c^{-1} is the inverse of the diagonal matrix H_c and could be easily calculated. δ_c^* and R_c^* are estimated as follows:

$$\delta_c^* = \sum_{i=1}^{N} y_{ic} \acute{\psi} \left(|\vec{a}_c| \right) \left[\left(\acute{\psi} \left(|\vec{a}_i| \right) \sum_{j=1}^{d+1} \frac{1}{\acute{\psi}(a_{cd})} \right) - 1 \right] \tag{21}$$

$$R_c^* = \frac{-1}{\sum_{i=1}^{N} y_{ic}} \left(\frac{1}{\acute{\psi} \left(a_{c1} \right)}, \ldots, \frac{1}{\acute{\psi} \left(a_{cD+1} \right)} \right) \tag{22}$$

Once M_c^{-1} and D_c are estimated, we can now implement the iterative formula of the Newton-Raphson algorithm as expressed by (11). Note that this algorithm requires starting values for $\{\hat{\vec{a}}_c\}^{(0)}$. In our implementation, we have used the method of moments estimators of the Dirichlet distribution [14] to define these initial values as follows:

$$\hat{a}_{cj}^{(0)} = \frac{\left(\acute{p}_{11} - \acute{p}_{21} \right) \acute{p}_{1j}}{\acute{p}_{21} - \left(\acute{p}_{11} \right)^2}; \;\; j = 1, \ldots, d; \;\; c = 1, \ldots, C \tag{23}$$

$$\hat{a}_{cd+1}^{(0)} = \frac{\left(\acute{p}_{11} - \acute{p}_{21} \right) \left(1 - \sum_{j=1}^{d} \acute{p}_{1j} \right)}{\acute{p}_{21} - \left(\acute{p}_{11} \right)^2} \tag{24}$$

where $\acute{p}_{1j}$ and $\acute{p}_{21}$ are given as follow:

$$\acute{p}_{1j} = \frac{1}{N} \sum_{i=1}^{N} x_{ij}; \;\; j = 1, \ldots, D+1 \tag{25}$$

$$\acute{p}_{21} = \frac{1}{N} \sum_{i=1}^{N} x_{i1}^2 \tag{26}$$

Algorithm 1: EM algorithm for Dirichlet mixture model

Input : $\{\vec{X}_i\}_{(i=1,\dots,N)}$, M
Output: $\{\hat{\vec{\mathbf{P}}}\} = \left[\hat{\vec{\alpha}}_1, \dots, \hat{\vec{\alpha}}_C, \hat{\vec{a}}_1, \dots, \hat{\vec{a}}_C\right]$
begin
  Initialization
  Apply the K-means algorithm to cluster the data set $\{\vec{X}_i\}$ into C components;
  Estimate the initial set of parameters of each component using (23) and (24);
  repeat
    Expectation
    Estimate $\{\hat{y}_{ic}\}$ $(i = 1, \dots, N; c = 1, \dots, C)$ using (7);
    Maximization
    Estimate $\left[\hat{\alpha}_c\right]$ $(c = 1, \dots, C)$ using (8);
    Estimate $\left[\hat{a}_{cj}\right]$ $(c = 1, \dots, C; j = 1, \dots, d)$ using (11);
  until *the change in (6) is negligible*;
  Return $\hat{\mathbf{P}}$;
end

The Newthon-Raphson algorithm converges, as our estimation of a_{cj} changes by less than a small positive value ϵ with each successful iteration, to $\widehat{a}_{cj}$.

The EM algorithm can now be used to estimate the maximum likelihood of the distribution parameters. Note that EM is highly dependent on initialization [20]. To alleviate this problem, a common solution is to perform initialization by mean of clustering algorithms. For this purpose we first implement the K-means algorithm in order to partition the set $\{\vec{X}_i\}_{i=1\dots,N}$ into C components. Then, based on such partition, we estimate the parameters of each component using the method of moment estimator of the Dirichlet distribution and set them as initial parameters to the EM algorithm. The steps of EM algorithm for Dirichlet mixture model is summarized in Algorithm 1.

2.4 *Estimating the Number of Components in the Mixture*

The use of the Dirichlet mixture model allows us to give a flexible model to describe the users' feature vectors. To form such a model, we need to estimate C, the number of components and the parameters for each component. First, the number of components C is an unknown parameter that must be estimated. Several model selection approaches have been proposed to estimate C [21, 22]. In this paper, we implemented a deterministic approach that uses the EM algorithm in order to obtain a set of candidate models for the range value of the C (from 1 to $C_{\max}$, the maximal number of components in the mixture) which is assumed to contain the optimal C [20]. We employ the integrated classification likelihood Bayesian information criterion (*ICL-BIC*) which is one the powerful methods to identify the

Algorithm 2: Estimating the number of components in the mixture

Input : $\{\vec{X}_i\}_{(i=1,\ldots,N)}$, $Cmax$
Output: The optimal number of components $\hat{C}$
begin
  for $C = 1$ **to** $Cmax$ **do**
    if *C==1* **then**
      Estimate $\{\vec{a}\}$ based on (11);
      Compute the value of ICL-BIC(C) using (27);
    else
      Estimate the parameters of the mixture using Algorithm 1;
      Compute the value of ICL-BIC(C) using (27);
    end
  end
  Select $\hat{C}$, such that $\hat{C} = arg\ min_C \Big\{\text{ICL-BIC}(C),\ C = 1, \ldots, Cmax\Big\}$;
end

correct number of clusters in the context of multivariate mixtures such as Dirichlet even when the component densities are misspecified [23]. *ICL-BIC* is given by:

$$ICL - BIC\,(c) = -2\ log(L_c) + N_p\ log(N) - 2\sum_{i=1}^{N}\sum_{c=1}^{C}\hat{y}_{ic}\ log\left(\hat{y}_{ic}\right) \quad (27)$$

where (L_c) is the logarithm of the likelihood at the maximum likelihood solution for the investigated mixture model N_p is the number of parameters estimated. The number of components that minimize ICL-BIC(C) is considered as the optimal value of the M. The procedure of estimating the number of components in the mixture is summarized in Algorithm 2.

2.5 *Automatic Identification of Anomalous Nodes*

Once the optimal number of components have been identified, we can use the results of the EM algorithm in order to derive a classification decision about the membership of $\vec{X}_i$ to each component in the mixture. In fact, the EM algorithm yields the final estimated posterior probability $\hat{y}_{ic}$, the value of which represents the posterior probability that $\vec{X}_i$ belongs to component c. We assign $\vec{X}_i$ to the component that corresponds to the maximum value of $\hat{y}_{ic}$. We thus divide the set of node feature vectors into several components. As discussed earlier, in our approach anomalous nodes are characterized by small feature values. Therefore, we are interested by the Dirichlet component which contains vectors with the lowest values. To identify such a component, we first compute, for each component in the mixture, the average of the projected feature values along each attribute.

Algorithm 3: Automatic identification of Anomalous nodes

Input : A network of N nodes
Output: A set A of anomalous nodes
begin
  For a given network, estimate a neighborhood cohesiveness feature vector $\vec{X}_i$ for each node using (1);
  Normalize $\{\vec{X}_i\}$, in such way that the summation of all the d element of the vector $\vec{X}_i = (x_{i1}, \dots, x_{id})^T$ is smaller than one;
  Apply Algorithm 2 to group the feature vectors $\{\vec{X}_i\}$ into C Dirichlet components;
  Use the results of EM to decide about the membership of $\vec{X}_i$ in each component;
  Select the Dirichlet component that corresponds to the smallest feature values;
  Identify nodes associated with the set of $\vec{X}_i$ that belong to the selected Dirichlet component and store them in A;
  Return A;
end

Then, we select the component with the smallest average value as our target component. Accordingly, nodes associated with the set of $\vec{X}_i$ that belong to such a component correspond to anomalies. The steps described in Algorithm 3 have been implemented to automatically identify anomalous nodes.

Our approach is able to accurately identify all the anomalous nodes in the network depicted by Fig. 1. In the following, we empirically demonstrate the suitability of the proposed methods using more complex synthesized and real networks.

3 Experiments

Hereafter, we describe a series of experiments to analyze the behavior of the proposed approach. The analysis is performed with two main objectives: (1) The assessment of results quality through performance comparative analysis using synthesized networks and (2) Illustrating the suitability of the proposed approach on real networks.

3.1 Experiments on Synthesized Networks

In this section, we conduct two testbeds experiments. In the first set of experiment, we generated synthetized networks with variable size and we fixed percentage of outliers, which is, of course, propositional to the size of each network. In the second set of experiments, we fixed the size of the generated networks but we significantly varied the number of outliers in each network. In all experiments, We compare the performance of the proposed approach to that of OddBall [8] and SCAN [10]. It has been shown in pervious works that OddBall and SCAN handle effectively anomalous nodes. Details and results of our experiments are given in what follows.

Experiment 1 To evaluate the quality of results, we generated synthesized networks with (community structures and anomalies) ground truth. The networks were generated according to the planted partitions model [24]. Specifically, we generated five networks with a varied number of normal and anomalous nodes. For each network, we specify different densities of the background noise on which dense structures (communities) are planted with various densities which are uniformly sampled from [0.1, 0.9]. Normal nodes fall within community structures and form densely connected groups. Anomalous nodes are sparsely distributed and the density of their connections conforms to the specified background noise density, which is uniformly drawn from [0.0, 0.1].

For the purpose of illustration Fig. 2 depicts examples of two generated, small-sized, networks with 250 nodes (Fig. 2a) and 350 nodes (Fig. 2b). In these figures, colored nodes correspond to normal nodes that from community structures. The remaining nodes (that is, uncolored nodes), correspond to anomalous nodes that do not belong to any community. As we can see from Fig. 2, anomalous nodes are sparsely connected. Most of them are: nodes having many connections across communities, bridging nodes between communities, and peripheral nodes. The networks depicted by Fig. 2 are provided here just for the purpose of illustration as they are small sized and thus easy to visualize. The generated networks used for the evaluation, however, are more complex. In the following, we describe the networks used in the experiments and the results of compared algorithms.

Table 1 summarizes the main characteristics of the five generated networks. As can be seen from Table 1, the network size varies from 1000 to 10,000 nodes, while the number of anomalies varies from 50 to 500. Our goal is to evaluate the robustness of competing algorithms with an increasing number of anomalous nodes. To this end, we used the following standard metrics: (1) Correct Detection Rate (CDR), measuring the proportion of anomalous nodes that are correctly identified as anomalous, (2) False Alarm Rate (FAR), corresponding to the proportion of normal nodes incorrectly flagged as anomalies, and (3) F-measure, corresponding to the harmonic mean between precision and recall of the anomalous nodes class.

We used our approach to identify anomalies in each of the generated networks. To this end, we set $C\max$ to 5[1] in all our experiments and then selected the number of components that minimize ICL-BIC. Interestingly, we found that the estimated node feature vectors are well fitted by two to three Dirichlet components. The component which contains vectors with the lowest values correspond to anomalous nodes. As previously discussed, to demonstrate the capability of our approach, we compared

[1]The reader should be aware that the value of $C\max$ is not limited to 5 and the user can choose other values.

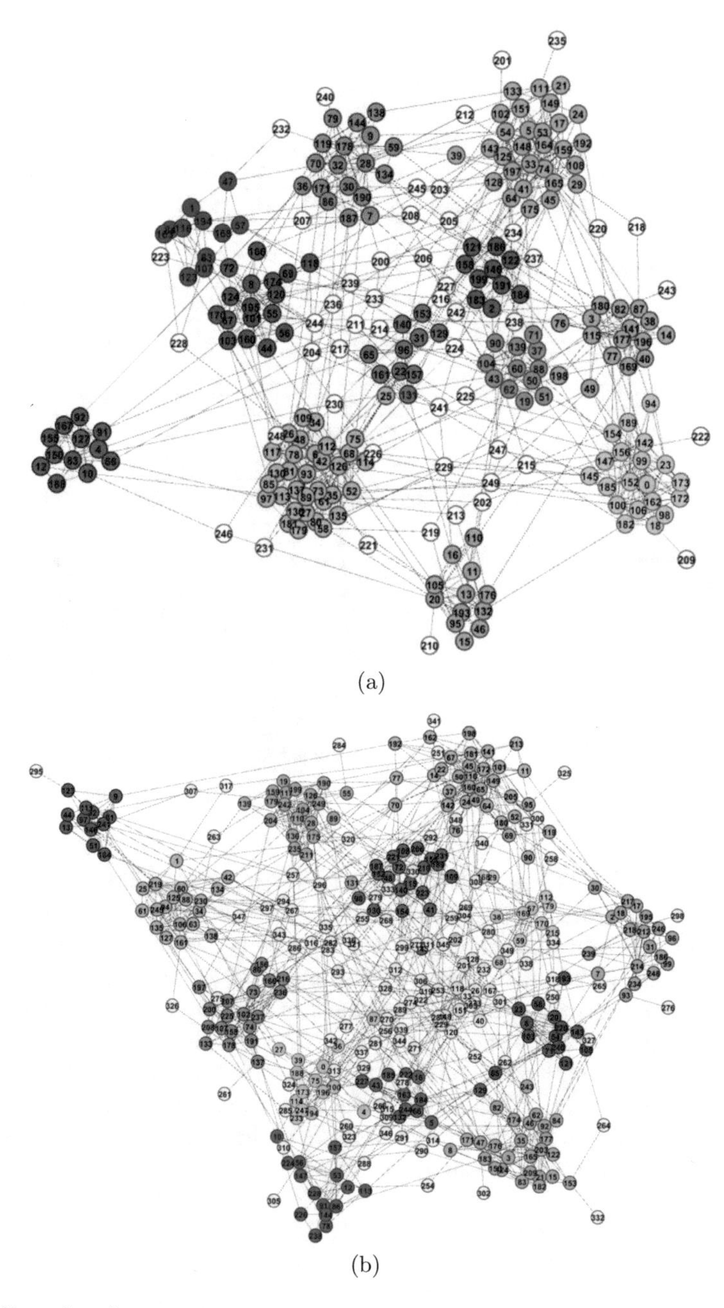

Fig. 2 Examples of generated networks with community structures (colored nodes) and anomalous nodes (uncolored nodes). As we can be seen, as the number of anomalous nodes increase, the more challenging to distinguish them from nodes forming community structures. (**a**) A network with 250 nodes. 50 nodes among them are anomalies. (**b**) A network with 350 nodes. 100 nodes among them are anomalies

Table 1 Synthesized network description

Network	#Nodes	#Normal nodes	#Anomalies
Network_1K	1000	950	50
Network_2K	2000	1900	100
Network_3K	3000	2850	150
Network_5K	5000	4750	250
Network_10K	10,000	9500	500

Table 2 F-measure (%)

Network	Proposed	OddBall	SCAN
Network_1K	92%	91.66%	**95.83%**
Network_2K	**95.52%**	91.20%	91.75%
Network_3K	**92.65%**	**92.65%**	91.94%
Network_5K	91.01%	**91.96%**	91.01%
Network_10K	**99.49%**	91.47%	97.95%
Average F-measure	**94.13%**	91.79%	93.70%

Bold values highlight the best results.

Table 3 False alarm rate—FAR (%)

Network	Proposed	OddBall	SCAN
Network_1K	0.63%	**0.21%**	0.63%
Network_2K	**0.42%**	0.47%	**0.42%**
Network_3K	0.66%	**0.83%**	0.42%
Network_5K	0.88%	0.42%	**0.12%**
Network_10K	0.31%	0.45%	**0.10%**
Average FAR	0.58%	0.39%	**0.34%**

Bold values highlight the best results.

it with OddBall and SCAN. OddBall[2] is a feature-based algorithm which provide a ranked list of nodes expecting anomalies to come first. OddBall need thus the target number of anomalies to be specified to formally distinguish between normal and anomalous nodes. In our experiments, and in order to compute CDR, FAR and F-measure, we simply set the real number of anomalies in the network to this algorithm. The second compared algorithm, SCAN, requires a similarity threshold as an input parameter. In our experiments, different runs were executed for different values of the threshold, and the best results were chosen. Performance results are summarized in Tables 2, 3 and 4. For the sake of discussion, all the values of the evaluation metrics in this table are shown in percentage (%).

[2]The implementation of OddBall is available from http://www.andrew.cmu.edu/user/lakoglu/#tools.

Table 4 Correct detection rate—CDR (%)

Network	Proposed	OddBall	SCAN
Network_1K	**95.83%**	91.66%	**95.83%**
Network_2K	**98.96%**	91.75%	91.28%
Network_3K	**97.31%**	92.61%	91.94%
Network_5K	**97.59%**	91.96%	85.26%
Network_10K	**99.59%**	91.56%	97.99%
Average CDR	**97.86%**	91.91%	92.46%

Bold values highlight the best results.

Overall, our approach displays a competitive performance as the number of anomalous nodes increase. As can be seen form Table 2, the proposed approach reports, on average, the highest F-measure (94.13%) compared to OddBall (91.79%) and SCAN (93.70%). In terms of FAR (see Table 3), the average performance of the compared algorithms is quite comparable. SCAN reports the best average FAR (0.34%), followed by OddBall (0.39%) and our approach (0.58%). All approaches report an acceptable low FAR reflecting the fact that only a very low portion of normal nodes that are erroneously classified as anomalies. Finally, we note that for all the generated networks, a certain number of anomalous nodes were flagged as normal by all the competing algorithms. In fact, as can be seen for Table 4, our approach, OddBall and SCAN report an average 97.86%, 91.91% and 92.46% CDR respectively. This means that 2.14%, on average, of anomalous nodes were incorrectly flagged as normal by our approach, while 8.09% and 7.54%, on average, of anomalous nodes were also misclassified as normal by OddBall and SCAN respectively.

To summarize, the results presented in Tables 2, 3 and 4 suggest that all the compared algorithms provide meaningful results. Our approach, however, achieve the best results especially in term of CDR and F-measures with a very low FAR which is less that 1% (the highest value reported by our approach is 0.88% on Network_5K).

Experiment 2 In the previous experiments, the amount of anomalous nodes was fixed to 5% of the network size. In this set of experiments, we fixed the size of the network instead and we varied the percentage of anomalous nodes. Specifically, we generated four networks containing 4000 nodes. In each network, the percentage of anomalies varies from 5 to 20% of the network size. Figure 3 illustrates the results of the competing algorithms, evaluated with F-measure, FAR and CDR.

As we can see from Fig. 3, the proposed approach displays consistent performance and is less affected by the increasing percentage of anomalous nodes in the investigated networks. Overall, compared to OddBall and SCAN, our method withstands the increasing value of the percentage of anomalies and maintain a good detection accuracy. In our experiments, we found that the proposed approach identifies most of anomalous node (which is reflected by the high CDR values as depicted by Fig. 3c), at the expense of also of misclassifying a very few number of normal nodes as anomalies (which is reflected by the lowest FAR values as depicted by Fig. 3b). In addition to the good results achieved by the proposed

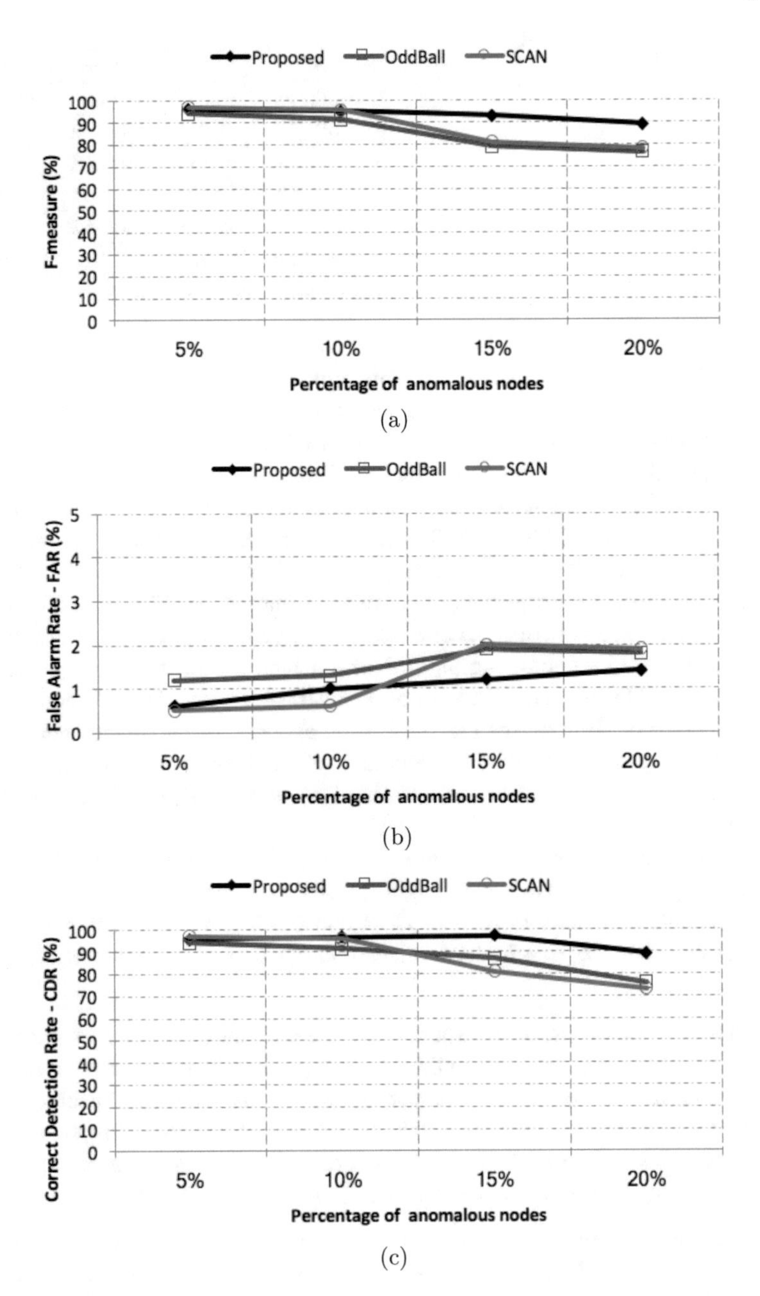

Fig. 3 Accuracies of compared algorithms as a function of increasing percentage of anomalous nodes. (**a**) F-measure (%). (**b**) False alarm rate—FAR (%). (**c**) Correct detection rate—CDR (%)

method, compared to OddBall and SCAN, our approach has the advantage to perform anomaly detection in an automatic fashion without asking the user to set the target number of anomalies to be selected from a ranked list (like OddBall) or requiring a similarity threshold to be set (which is the case of SCAN). In real application, it is hard to set an accurate value of these parameters since, in contrast to synthesized networks, we don't have, in the vast majority of cases, any prior knowledge about the network under investigation.

3.2 Application of the Proposed Approach on Real Networks

In this subsection, we put our approach to work using real networks. Here, it is worth noting that, as observed in several previous studies [3, 5], there is a shortage of ground truth real network in the anomaly detection scenarios. In this setting, comparing and evaluating network-based anomaly detection approaches on real networks becomes a challenging task since the evaluation should be done via an external knowledge source (domain expert) which is not always available. In the following, we saliently illustrate the suitability of the proposed approaches on four real networks. Since the available real networks are unlabeled, we apply our approach in an exploratory fashion and report our finding. Note that we did not considered OddBall and SCAN for comparison due to: (1) the absence of ground truth which makes an objective comparison not obvious and (2) it is hard to set an appropriate value of their input parameters. For OddBall, we have to specify how many nodes should we consider as anomalies from the ranked list. This information is not obvious to specify as we don't know where to stop reading the ranked list, especially in real scenarios. For SCAN, we have to set a similarity threshold. With real networks, an appropriate value for this threshold is difficult to identify as it differs from one network to another since there is no single threshold suitable for all networks.

Table 5 shows the main characteristics of the four real networks considered in our experiments. As we can see from this table, the size of the network varies from 36,692 to 81,306 nodes, while the number of edges varies from 106,978 to 1,768,149. The DBLP data set[3] contains the co-publications of authors. For each paper published, it gives several information such as the paper title, the authors, the year and the publication venue. We built a co-authorship network from year 2015. The remaining three networks, that is, Enron, Epinions and Twitter are downloaded from the Stanford Large Network Dataset Collection repository.[4] In nutshell, Enron, is a communication network reflecting email exchanges between Enron employees. Epinions, is a who-trust-whom online social network of a general consumer review

[3]https://aminer.org/citation.

[4]http://snap.stanford.edu/data/index.html.

Table 5 Real network description

Network	#Nodes	#Edges
DBLP	47,917	106,978
Enron	36,692	183,831
Epinions	75,879	508,837
Twitter	81,306	1,768,149

site Epinions.com. Finally, the Twitter data is a social network reflecting social circles from Twitter.

We applied our approach to the above networks. Interestingly, we found that the estimated nodes' feature vectors are well fitted by three Dirichlet components. The component contacting the smallest feature values, represent atypical, sparsely distributed nodes. Specifically, for DBLP, we found 8359 nodes out of 47,917, that is, 17.44% of the total number of nodes, correspond to anomalies. Among the 8359 identified anomalous nodes, 4789 nodes, that is, 57.29%, have all the four elements of their associated feature vectors equal to zero. This highlight the sparseness of the connections of the identified anomalous node across the DBLP network. We found that identified anomalous nodes mainly reflect occasional or unique co-publication between authors in contrast densely connected nodes (normal nodes) which reveal strong collaboration. Specifically, we found that most of the identified anomalies form many isolated chains of at least two connected nodes. We found also anomalous nodes forming isolated circles (with no cliques/triangles) and (near) star structures. These atypical connections correspond to anomalous nodes topological structures observed in previous studies [3, 8].

We also observed the same phenomenon in the remaining three networks, that is, the presence of many isolated chains of nodes, isolated circles and stars. Statistically speaking, in Enron we found 14,475 nodes out of 36,692, that is, 39.45% of the total number of nodes, correspond to anomalies. Among the 14,475 identified anomalous nodes, 12,240 nodes, that is, 84.55%, have all the four elements of their associated feature vectors equal to zero. The number of sparsely connected node identified by our approach in Epinion is 65,385. 98.19% among them have all their feature values equal to zero. This means that 64,207 nodes, among 65,385 anomalies, did not share at least one connection with their immediate neighbor. This clearly indicates the high sparseness of connections in the who-trust-whom Epinion network. Finally, in Twitter, we identified 27,267 anomalous nodes out of 81,306, that is, the proportion of sparsely connected nodes is 33.53%. Out of the 27,267 identified nodes 18.47% of them did not share any neighbors (all the elements of their feature vectors equal to zero). Overall, the proportion of anomalous nodes identified by our approach appears to be relatively high. We claim that such results are not surprising as large real networks are very sparse. A general perception in the field is that the number of anomalies in a data set is likely to be very small. This is true for some applications. However, the networked data available nowadays are usually sparse and rich of atypical and sparsely connected nodes that may reveal interesting social phenomenon and behavior.

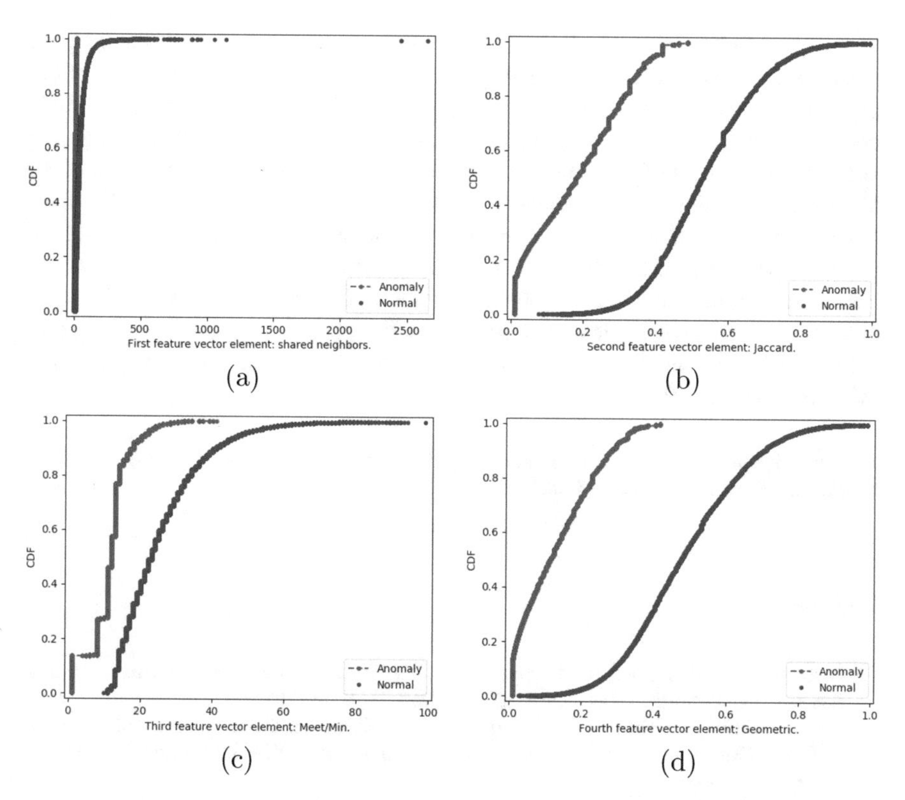

Fig. 4 The cumulative distribution function for the Twitter network nodes' features. (**a**) Shared neighbors. (**b**) Jaccard. (**c**) Meet/Min. (**d**) Geometric

For the purpose of illustration, Fig. 4 shows the cumulative distribution function (CDF) for the four feature values (as estimated by (1)) of the Twitter network. The plots in Fig. 4 are very representative of the general behavior of our approach on DBLP, Enron and Epinions. The knowledgeable reader can observe from this rendering that there is a clear distinction between anomalous and normal nodes. In contrast to anomalies, the curve for normal nodes is much more skewed towards large number. In fact, in contrast to anomalies, normal nodes tend to exhibit a high level of close interactions within the network.

4 Conclusion

In this paper, we devised a principled approach for automatic identification of anomalous nodes in networks. The experiments on synthesized networks illustrate that our approach identify anomalies with a high accuracy and resist well to the increasing number of anomalous nodes. Experiments on real networks suggest that the proposed method is able to handle sparse large networks characterized by the presence of many atypical and sparsely connected nodes. Recall that the aim of this paper is identifying nodes that are different or deviate significantly from closely connected nodes in the graph. So, in our experiments on real networks, we have not provided a detailed analysis of the social meaning and the specific roles of the identified anomalous nodes which is itself a very important issue, but, however, out of the scope of this study. Our approach identifies anomalous nodes based on the topological structure of the network. Although such identified nodes are considered anomalous from the topological point of view, we believe that they may characterize some social phenomenon that can be evaluated, by the domain experts, using existing knowledge or newly suggested hypotheses. Further investigations in this direction are called for.

Acknowledgements This work is supported by research grants from the Natural Sciences and Engineering Research Council of Canada.

References

1. O. Boutemine, M. Bouguessa, Mining community structures in multidimensional networks. Assoc. Comput. Mach. Trans. Knowl. Discov. Data **11**(4), 51 (2017)
2. J. Li, K. Cheng, L. Wu, H. Liu, Streaming link prediction on dynamic attributed networks, in *Proceedings of the 11th ACM International Conference on Web Search and Data Mining* (2018), pp. 369–377
3. L. Akoglu, H. Tong, D. Koutra, Graph based anomaly detection and description: a survey. Data Min. Knowl. Disc. **29**(3), 626–688 (2015)
4. E.G. Tajeuna, M. Bouguessa, S. Wang, Modeling and predicting community structure changes in time evolving social networks, in *IEEE Transactions on Knowledge and Data Engineering* (IEEE, Piscataway, 2018). https://doi.org/10.1109/TKDE.2018.2851586
5. C.C. Aggarwal, Outlier detection in graphs and networks, in *Outlier Analysis* (Springer, Heidelberg, 2013), pp. 343–371
6. D. Savage, X. Zhang, X. Yu, P. Chou, Q. Wang, Anomaly detection in online social networks. Soc. Networks **39**, 62–70 (2014)
7. V. Chandola, A. Banerjee, V. Kumar, Anomaly detection: a survey. Assoc. Comput. Mach. Comput. Surv. **41**(3), 15 (2009)
8. L. Akoglu, M. McGlohon, C. Faloutsos, Oddball: spotting anomalies in weighted graphs, in *Advances in Knowledge Discovery and Data Mining* (2010), pp. 410–421
9. M.M. Breunig, H.-P. Kriegel, R.T. Ng, J. Sander, Lof: identifying density-based local outliers, in *ACM SIGMOD International Conference on Management of Data* (2000), pp. 93–104
10. X. Xu, N. Yuruk, Z. Feng, T.A. Schweiger, Scan: a structural clustering algorithm for networks, in *Proceedings of the 13th ACM SIGKDD International Conference on Knowledge Discovery and Data Mining* (2007), pp. 824–833
11. H. Sun, J. Huang, J. Han, H. Deng, P. Zhao, B. Feng, gSkeletonClu: density-based network clustering via structure-connected tree division or agglomeration, in *IEEE International Conference on Data Mining* (2010), pp. 481–490
12. H. Tong, C.-Y. Lin, Non-negative residual matrix factorization with application to graph anomaly detection, in *Proceedings of the SIAM International Conference on Data Mining* (2011), pp. 143–153

13. T.H. Haveliwala, Topic-sensitive PageRank: a context-sensitive ranking algorithm for web search. IEEE Trans. Knowl. Data Eng. **15**(4), 84–796 (2003)
14. N. Bouguila, D. Ziou, J. Vaillancourt, Unsupervised learning of a finite mixture model based on the Dirichlet distribution and its application. IEEE Trans. Image Process. **13**(11), 1533–1543 (2004)
15. D.S. Goldberg, F.P. Roth, Assessing experimentally derived interactions in a small world. Proc. Natl. Acad. Sci. **100**(8), 4372–4376 (2003)
16. Z. Ma, *Non-Gaussian Statistical Models and Their Applications* (KTH-Royal Institute of Technology, Stockholm, 2011). PhD thesis
17. S. Boutemedjet, D. Ziou, N. Bouguila, Model-based subspace clustering of non-gaussian data. Neurocomputing **73**(10), 1730–1739 (2010)
18. T. Bdiri, N. Bouguila, Positive vectors clustering using inverted Dirichlet finite mixture models. Expert Syst. Appl. **39**(2), 1869–1882 (2012)
19. F.A. Graybill, *Matrices with Applications in Statistics* (Wadsworth International Group, Belmont, 1983).
20. M.A.T. Figueiredo, A.K. Jain, Unsupervised learning of finite mixture models. IEEE Trans. Pattern Anal. Mach. Intell. **24**(3), 381–396 (2002)
21. P. Smyth, Model selection for probabilistic clustering using cross-validated likelihood. Stat. Comput. **10**(1), 63–72 (2000)
22. M. Bouguessa, S. Wang, H. Sun, An objective approach to cluster validation. Pattern Recogn. Lett. **27**(13), 1419–1430 (2006)
23. D. Peel, G.J. McLachlan, Robust mixture modelling using the t distribution. Stat. Comput. **10**(4), 339–348 (2000)
24. A. Condon, R.M. Karp, Algorithms for graph partitioning on the planted partition model. Random Struct. Algoritm. **18**(2), 116–140 (2001)

Printed in the United States
By Bookmasters